THE SYSTEMS APPROACH

THE SYSTEMS APPROACH

by C. West Churchman

A Delta Book

A DELTA BOOK

Published by
DELL PUBLISHING CO., INC.
1 Dag Hammarskjold Plaza
New York, N.Y. 10017

Delta ® TM 755118, Dell Publishing Co., Inc.
Reprinted by arrangement with
Delacorte Press, New York, New York
Manufactured in The United States of America
Seventeenth Printing

ACKNOWLEDGMENTS

This book was written while the author was conducting social science research on the managerial problems of research and development, supported by the National Aeronautics and Space Administration. Many of the ideas were inspired by this research effort and by the discussions of the social-science seminar of the Space Sciences Laboratory at the University of California, Berkeley.

CONTENTS

PREFACE

There is no question that in our age there is a good deal of turmoil about the manner in which our society is run. Probably at no prior point in the history of man has there been so much discussion about the rights and wrongs of the policy makers, whether they be the politicians in Albany or Sacramento, in Washington, Paris, or Moscow, the managers of far-flung industrial firms, or the people who run educational institutions. In all cases the citizen feels a perfect right to have his say about the way in which the managers manage.

Not only has the citizen become far more vocal, but he has also in many instances begun to suspect that the people who make the major decisions that affect our lives don't know what they are doing. They don't know what they are doing simply because they have no adequate basis to judge the effects of their decisions.

To many it must seem that we live in an age of moronic decision making. About all that the decision maker can do is pick on one aspect of the situation and push that as hard as possible, arguing against his enemies on the basis that they are failing to sense the true situation.

Thus in the case of the war in Vietnam the hawks latch on to the Communist threat and let it go at that. They keep pushing this one aspect of the situation and ignore all else that their opponents are saying. The doves, on the other hand, are equally bad. They keep saying that the war is a miserable failure, that we ought to pull out of it, and that we had no business getting into it in the first place. In the midst of the more or less violent debate there are a few groups who are playing it cool. Among these can be counted the so-called scientific mind that attempts to stand apart, to look at what is going on, and to make a judgment about all of the ramifications of the system as far as it can see.

The idea that the dispassionate and yet clear mind of the scientist can aid in decision making is an old-fashioned one. Plato had it many years ago when he thought he could begin to design the underlying model of a city-state, as he did in his *Republic*. Down the ages, every so often a writer has set down what in his opinion are the essential ingredients in a messy situation in order to untangle the various factors and set the matter straight in a scientific and objective fashion.

A very fascinating development of this kind occurred in World War II, when the British Admiralty asked teams of scientists to consider some of the pressing problems that Britain faced during the first bombings by the Nazis. What is especially interesting about this story is that the scientists kept asking stupid questions. For example, the British were having a good deal of trouble knocking out the German submarines in the English Channel. The scientists noted that the depth charges dropped from aircraft were set so that the charge did not go off until at least 35 feet below the surface. The scientists asked the stupid question: Why not try to set the charges so they go off at a shallower depth? Once you've asked a stupid question, then you have to defend your right to ask it, and the scientists pointed out some of the weaknesses in the assumptions that were made by the military in the manner in which the aircraft was approaching its target. Eventually some experiments were run, and sure enough, the submarine kill went up significantly as a result of setting the charges at a shallower depth.

The story is illustrative of a theme that will run throughout this book, namely, that when one is considering systems it's always wise to raise questions about the most obvious and simple assumptions.

The success of the scientific teams in the military in World War II was outstanding both in Great Britain and in America. As a consequence after the war there was a rush to apply the same kind of thinking, which then was called "operations research," to various nonmilitary problems, and in particular to industry. At first the problems considered were rather small. The scientists studied production as well as some minor problems of marketing and finance. A few sporadic attempts were made to solve some problems in transportation, e.g., the design of roadways. But luckily along came a computer that was an enormous aid to the scientist. At first the computer itself was used in a relatively insignificant role, e.g., in tasks like bookkeeping. Then people began to see that computer capability as it was opening up in the 1950s suggested the possibility of using the computer as a way of processing enormous amounts of information. Thus came the idea for the SAGE system, a system that would supply to the defense agencies of the United States information about the location of our own aircraft and of all enemy aircraft, and information about suspicious objects in the air. Information was processed and displayed to the managers of the system in a very precise and intelligible fashion. The computer had started to grow up. Along with this growth in computer capability came a widening of the scientist's interests in policy problems. One outstanding result was the creation of RAND and similar nonprofit corporations designed to study some of the important strategic as well as tactical problems of the military.

As the scientist's perspective widened, he began to think of his approach as the "systems approach." He saw that what he was chiefly interested in was characterizing the nature of the system in such a way that the decision making could take place in a logical and coherent fashion and that none of the fallacies of narrow-minded thinking would occur. Furthermore, using his scientific knowledge he expected to be able to develop measures which would give as adequate information as possible about the performance of the system.

In time the decision makers, both in industry and in government, began to see the possible values of the so-called systems approach. For example, Governor Brown of California in the early 1960s proposed to the aerospace industry that they submit to him four systems science designs for crime, sanitation, information and transportation. Governor Brown's idea was that people who are well trained in the design of complex hardware systems, such as space missiles, should be able to apply their thinking to the critical decision-making problems of the state. Mayor Lindsay has recently set up a Management Science Council to assist the City of New York in some of its critical decision problems. Governor Rockefeller of the State of New York has indicated his intention to use operations research. In practically every office of the government there are operations researchers, management scientists, system scientists, all attempting to look at the problems of the United States government from the so-called systems approach.

This book is an attempt to examine what the "systems approach" means. It does so not from the point of view of "selling" the idea, but rather by examining its validity in the climate of a debate. A great deal of nonsense has been written about the systems approach, because once an idea becomes popular it can be sold, and naturally some of its sellers are out to make a profit.

Now, it is sheer nonsense to expect that any human being has yet been able to attain such insight into the problems of society that he can really identify *the* central problems and determine how they should be solved. The systems in which we live are far too complicated as yet for our intellectual powers and technology to understand. Given the limited scope of our capability to solve the social problems we face, we have every right to question whether any approach—systems approach, humanist approach, artist's approach, engineering approach, religious approach, psychoanalytic approach—is *the* correct approach to the understanding of our society. But a great deal can be learned by allowing a clear statement of an approach to be made in order that its opponents may therefore state their opposition in as cogent a fashion as possible.

At the end of the book I conclude that *the* systems approach really consists of a continuing debate between various attitudes of mind with respect to society. This of course is not a novel conclusion, but I hope it will assist the reader in formulating the kinds of question that should be asked at the present time about the various ways in which science can assist society in its decision making.

I. What is a system?

1. THINKING

Suppose we begin by listing the problems of the world today that *in principle* can be solved by modern technology.

In principle, we have the technological capability of adequately feeding, sheltering and clothing every inhabitant of the world.

In principle, we have the technological capability of providing adequate medical care for every inhabitant of the world.

In principle, we have the technological capability of providing sufficient education for every inhabitant of the world for him to enjoy a mature intellectual life.

In principle, we have the technological capability of outlawing warfare and of instituting social sanctions that will prevent the outbreak of illegal war.

In principle, we have the capability of creating in all societies a freedom of opinion and a freedom of action that will minimize the illegitimate constraints imposed by the society on the individual.

In principle, we have the capability of developing new technologies that will release new sources of energy and power to take care of physical and economic emergencies throughout the world.

In principle, we have the capability of organizing the societies of the world today to bring into existence well-developed plans for solving the problems of poverty, health, education, war, human freedom and the development of new resources.

If the human being has the capability of doing all of these things, why doesn't he do it? Is there some perverse streak that runs throughout the human race that makes one human being indifferent to the plight of another? Are we essentially faced with a type of moral degradation that permits us to ignore our neighbor for the sake of our own good?

Or is there some deeper and more subtle reason why, despite our enormous technological capability, we are still in no position to solve the major problems of the world? If we look over the list of problems, one aspect of them becomes quite obvious: these problems are interconnected and overlapping. The solution of one clearly has a great deal to do with the solution of another.

They are so interconnected and overlapping, in fact, that it is not clear at all where we ought to begin. For example, suppose we have made up our mind that the first problem to be solved is that of adequately feeding, sheltering, and clothing every inhabitant of the world. How should we begin to solve this problem? The technological capability is there. We can produce the food necessary to do the job and the housing materials that will provide shelter and the textiles that will clothe each person. Then why don't we do so? The answer is that we are not organized to do so. In other words, the last objective, the development of a set of organizations that will solve the major problems of the world, has to be handled first. Is this where we should begin? Why don't we simply organize the world to do the job of feeding, sheltering, and clothing? For example, why doesn't the United States pull together a conference at some peaceful spot in the world for the purpose of laying out and implementing a plan for feeding, sheltering and clothing the world's inhabitants?

The answer to this last question is that the United States is not sufficiently trusted to be the initiator of such a conference. Many nations of the world fear the military strength of the United States. The United States is capable of carrying on warfare in its own defense whenever it thinks it desirable to do so.

This means that another problem, namely, the problem of the insecurity of the world in the face of sanctioned warfare, has to be solved first. We have to create a world in which the nations of the world trust each other in somewhat the same way as the states of the United States trust each other. Therefore, the first problem to be solved is one of creating an international politics which would provide the environment for a worldwide conference on the solutions of the problems of worldwide poverty.

How can we create a satisfactory worldwide politics when such a great proportion of the human race is uneducated and therefore unaware of the fundamental problems of the world and their relationship to them? Distrust always occurs in the environment of ignorance. One cannot expect to create an enlightened international politics without also creating the educational background of each individual who will have something to say about how the world is run. Therefore, the first problem to be solved is the educational problem. How can we educate the peoples of the world so that the threat of ignorance is removed?

But clearly there is no adequate way of educating a starving man. Adequate education is based on the premise that the person being educated is himself adequately fed, sheltered, and clothed and is also in a satisfactory mental and physical state from the point of view of his health. Therefore, the first problems to be solved are the problems of health and poverty. And there we are, back at the beginning again.

We seem in fact to be faced with a dilemma. On the one hand, it would be extremely foolish to ignore the problems of the world today and, so to speak, bury our heads in our own pile of gold. On the other hand, there seems to be no adequate way even to think about the major problems of the world in any realistic sense.

But it may be that the real trouble here is in the way we started to think about the problems. We started to think about the problem by listing all the things we might in principle accomplish by our marvelous technology. After we had written the list, we asked ourselves where we should begin: with poverty, with health, with education, or where? Perhaps our trouble was that we didn't begin to think soon enough. The list we made came off the top of our heads without any prior thinking on our part other than a recollection of articles, books and speeches about these problems. Consequently, we looked for a beginning in a list that we had generated without thinking very much at all.

Now, logicians tell us that when we want to solve problems we should begin with the thinking process. Otherwise, we are apt to go off on entirely the wrong pathway in our exploration, and our thinking will come in far too late. It is much as though a man who is somewhat lost dashes off on the first pathway he sees and lets his feet carry him some distance before he begins to think in some logical way about how he should get himself out of his difficulty; but then it may be too late.

Suppose we contrast our thinking about the world's problems with a far more specific set of thoughts about the development of one type of technology, say, the development of a rocket that is capable of sending an object to the moon. Here we have the very specific objective of landing of an object on the moon within prescribed budgetary limits. We can start our thinking with the *central objective* and then begin to ask ourselves for a list of subobjectives which are obviously required in order to accomplish the central objective.

If we wish to land an object on the moon, then clearly we need (1) a propellant system, i.e., a substance capable of pushing the object out of the gravitational field of the earth; (2) the design of the "bird" that will fly to the moon and the boosters that will permit it to make its flight and to land satisfactorily. Quite obviously we will also need to have (3) a communication-control subsystem that will enable people on the earth to know where the object is and if necessary to control its flight, and to learn when it has landed. And, if the bird is to have a human inhabitant, then quite obviously we will have (4) to select and train the one or more people who are going to fly on it.

This "shopping list" of items that would be required in order to accomplish the purpose of landing something on the moon is not quite sufficient. We know we are going to have to ask a group of people to develop the propellant subsystem, the design of the bird and the booster subsystems, the communication and control subsystem, and the astronaut subsystem. We need to create guidelines that will enable them to do their job well. In other words, for each subsystem we are going to need a measure of the performance of the subsystem and a desired level of performance which we can call a "standard" for the subsystem. That is, we are going to have to tell the engineer that we need a propellant capable of lifting a certain weight and shape of the bird out of the gravitational field of the earth. And we are going to have to tell him the amount of money that we can spend on creating such a propellant. If we tell him these things clearly enough, then he will be capable of measuring the performance of a given propellant, and we will be capable of deciding whether or not the particular propellant that he offers comes up to the desired standard. This means we are going to have to be able to measure whether or not a subsystem conforms to the standard. If it does so, then we are in the position of accepting it and using it in the total system. If it does not, then we know we have further steps to take to develop a subsystem up to the level that we desire.

This is not all, however. We can't spend forever on the design of the subsystem. In fact, we can already begin to sense that if some of our efforts are delayed, then some of our other efforts are a waste of time. If it's going to take ten years to develop an adequate propellant, then we know we shouldn't develop highly trained astronauts to fly on the system next year because, by the time the propellant is fully developed, the astronauts will be too old to fly. Consequently we need a plan that will bring each subsystem up to the standard at a desired *time* so that the whole developmental effort goes along smoothly and there is no serious wastage on account of delay.

But because we can never make sure that a given set ot plans or aspirations will come to fruition, we need something else besides. We need to set down the explicit steps that we will be willing to take and capable of taking when the plans fail. This is perhaps one of the most neglected aspects of the system approach to design and planning. The planners are often far too optimistic about their success so that when failures occur they are in no position whatsoever to take the necessary steps because they have never thought about them before. In other words, to reiterate the point, *when you postpone thinking about something too long, then it may not be possible to think about it adequately at all.*

If finally, in the plan for the development of a system for placing an object on the moon, we throw in as a component the activities that determine the overall objective and the justification of each of the subsystems, the measures of performance and standards in terms of the overall objective, then the whole set of subsystems and their plans and their measures of performance constitute a "systems approach" to the problem of putting an object on the moon. This last component, which determines the overall objectives and relates the subsystem standards to the overall, can be called the "management subsystem." It is the subsystem that thinks about the overall plan and implements its thinking. If the management subsystem works correctly, its thinking goes on continuously. It thinks about the relationship of the overall objective to the components at the very outset. It does not postpone its thinking until a crisis is reached. It does not begin by listing a set of things it wants to do without bothering its head about the reason why it wants to do them. Each step of the plan is justified in terms of the overall objective. This does not mean that its thinking is rigid and closed because rigid and closed thinking is also inappropriate thinking. It thinks about how it should act when the unexpected occurs. Of course, the management subsystem may be surprised, because no thinking is ever perfect. But if the management subsystem is acting correctly, it is never caught in a situation in which prior thinking could have saved it.

Does this type of thinking about the whole system help very much in terms of our attitude toward the problems in the world today?

Well, at the outset we can notice one striking feature of the earlier discussion. In addition to listing a lot of problems that we would like to solve without thinking very much about the list, it's also clear that many of the problems on the list don't make much sense by themselves.

Do we really want to clothe, shelter, and feed every person of the world? That is, do we want to do this no matter what else happens? Nazi Germany developed an approach to adequate feeding, sheltering, and clothing. The first task, they said, was to get rid of all of the "undesirable" types of people in the society, e.g., Jews, and thereby reduce the total magnitude of the job as well as eliminate any opposition to the state's plans. Thus the Nazi state developed a subsystem that set to work to remove all of the socially undesirable elements, the elements that were mentally or physically unhealthy, or those that seemed to be a threat to the plan of the state.

The measure of performance of this subsystem is its ability to perform its elimination, and a well-laid-out plan for the state will be one in which a standard is set up for this subsystem. The subsystem would therefore fail if it did not meet the standard of eliminating various unsatisfactory people in the system. If this subsystem behaves properly and the subsystem for the creation of food, shelter, and clothing behaves properly, then the state can move toward its objective, and the overall objective of feeding, sheltering, and clothing will be accomplished in a highly efficient manner.

But we don't want to feed, shelter, and clothe the world in this manner. We want to feed, shelter, and clothe the world subject to conditions that create a free society. We don't believe that the way to solve the problems of mental and physical health is by the elimination of mentally and physically diseased.

What do we wish to accomplish? Can we actually state an objective that is as operationally clear as the objective of landing an object on the moon, subject to budgetary constraints? Or is it a foolish waste of time for us to think about the objectives of the inhabitants of the world in such terms?

Many people will recall that sometime during World War II the two great leaders of the English-speaking nations, Roosevelt and Churchill, met somewhere in the Atlantic Ocean and announced the "four freedoms." Although this announcement on the part of these leaders was undoubtedly inspirational, it was certainly lacking in any systems approach to the problems of the world simply because it failed to provide any way in which we could adequately think of how to begin. It failed to do this because it did not provide the statement of the objectives that would guide one in thinking about how to begin.

In this book we will be considering some ways of thinking about whole systems. We will begin quite modestly at first, not with the problems of the whole world, but with the problems of some very specific systems. Our chief interest will not be in hardware systems like the rocket to the moon, but rather in systems with humans in them. These are systems like industrial firms, hospitals, educational institutions, and so on.

All along, our effort will be to expand our capability of thinking about systems. Consequently, this is not a book on the subject of how each individual should learn to think; it is more concerned with the resources that a society has at its disposal for thinking better about its systems. Some of these resources can be described in terms of physical hardware, e.g., computer systems. Some of them can be described in terms of highly developed mathematical tools that assist managers in thinking about their systems. A book of this sort is an extension of the old-fashioned logic and rhetoric in which the student is trained adequately to think about the world. Some of Aristotle's lectures to his students consisted of a series of warnings about how they might fall into logical traps when they were confronted by various types of sophistry. Spinoza wrote one of his books on how to think, which he called *The Improvement of Human Understanding*. And in later days John Dewey enriched the world's literature with some very commonsensical prescriptions for the thinking process. But today our expanding technology provides us with all kinds of additional resources beyond the basic types of logic which were taught by Aristotle, Spinoza, Dewey, and others. There is a plethora of additional resources that we will want to explore as we develop some basic ideas on how to think in our century.

The idea of a "systems approach" is both quite popular and quite unpopular. It's popular because it sounds good to say that the whole system is being considered, but it's quite unpopular because it sounds either like a lot of nonsense or else downright dangerous—so much evil can be created under the guise of serving the whole.

Our best way to proceed will be to use the old tried and true method of debate. We'll let the systems enthusiast have his say, or rather, his say's, because we'll find that there are a number of definitions of the systems approach. His name will keep changing in accordance with the change of viewpoint: efficiency expert, management scientist, planner, and so on. His opponent will take on several guises, sometimes a doubting Thomas (who once questioned the grandest systems approach of all), sometimes an infuriated humanist. We won't expect the critic to be consistent, because strict adherence to consistency is itself a systems approach, but we will expect him to bring out clearly the doubts and evils he feels about too much "thinking."

To get a flavor of the enthusiast, listen to him for a bit.

"Is there something essential about the concept of a system as a way of thinking? There surely is. Systems are made up of sets of components that work together for the overall objective of the whole. The systems approach is simply a way of thinking about these total systems and their components. We have already seen one essential feature of this way of thinking, namely, that thinking enters in at the very outset in dictating the manner in which we describe what it is we are planning to do.

"We ought not to approach the world blindly, letting our observations and what other people tell us be the basis of our description. We shouldn't say that the world is made up of problems like poverty, health, education, and so on simply because these are the problems that everybody is talking about. We ought to ask ourselves at the very outset how to think about a large system, and our manner of thinking will dictate how we will describe the system. Some of the descriptions of systems are not obvious at all. There are ways of describing systems that would not occur to most people who tend to look at the world in one way, namely, the way that is most familiar to them. The systems approach will have to disturb typical mental processes and suggest some radical approaches to thinking. It may in fact be already quite radical for somebody to think first of all about the overall objective and then to begin to describe the system in terms of this overall objective.

"For example, if I ask you to describe an automobile, you may immediately switch off your thinking process and simply blurt out the things you recall about your own automobile—its wheels, engine, and shape. You start by saying, 'Well, an automobile is something that has four wheels and is driven by an engine.' I (in an attempt to switch on your thinking process) ask whether a three-wheeled automobile is a possibility. You have seen one and will readily admit this change in your description, still without thinking much about the meaning of the change. I, becoming more belligerent, pursue the matter further and ask you whether a two-wheeled automobile is a possibility. You begin to look puzzled, thus indicating that your thinking has been turned on at a low voltage. I go on, being cheerfully disagreeable, and ask you whether an automobile without any wheels whatsoever is also a possibility. You become more puzzled and think not about automobiles but about silly question posers. Yet to consider the wheel-less automobile is a creative way of looking at this system we call the automobile. It may be that the need for wheels is one of the major producers of traffic congestion and the inconvenience of the current automobile. An automobile that can float a few feet off the surface of the earth might provide a far more comfortable ride and produce far fewer problems of traffic congestion and even of accidents. And floating automobiles may be technically feasible in the future.

"The way to describe an automobile is *first* by thinking about what it is for, about its *function,* and not the list of items that make up its structure. If you begin by thinking about the function of the automobile, that is, what it is for, then you won't describe the automobile by talking about its four wheels, its engine, size, and so on. You will begin by thinking that an automobile is a mechanical means of transporting a few people from one place to another, at a certain prescribed cost. As soon as you begin to think in this manner, then your 'description' of the automobile begins to take on new and often quite radical aspects. That's the systems approach to automotive transportation.

"Or look again at the questions posed at the beginning of the chapter, the problems of the world today. From the systems point of view, we have to admit to ourselves that we may have begun incorrectly because we began by describing the world in terms of its structure, not its purpose; we began by talking about the world's inhabitants and the various defects of their environment. The 'world' that we were describing may not be what the 'world' will be in systems terms. As we begin to learn some lessons of systems thinking, perhaps we'll be able to end up with somewhat radical ways of thinking about the meaning of the world."

Now of course, all this sounds quite reasonable, as it has to a great many people. The differences arise when we try to make these ideas much more specific and applicable. Then we find that there are several systems approaches, not one. In this book we shall examine four different ideas as to what really constitutes *the* systems approach, and we shall juxtapose them in the context of a debate.

The debaters are these: (1) The advocates of *efficiency;* they claim that the best approach to a system is to identify the trouble spots, and especially the places where there is waste, e.g., unnecessarily high costs, and then proceed to remove the inefficiency. (2) The advocates of the use of *science* in approaching a system; they claim that there is an objective way to look at a system and to build a "model" of the system that describes how it works. The science that is used is sometimes mathematics, sometimes economics, sometimes "behavioral" (e.g., psychology and sociology). (3) The advocates of the use of human feelings, i.e., the *humanists;* they claim that systems are people, and the fundamental approach to systems consists of first looking at the human values: freedom, dignity, privacy. Above all, they say, the systems approach should avoid imposing plans, i.e., intervention of any kind. (4) The *anti-planners,* who believe that any attempt to lay out specific and "rational" plans is either foolish or dangerous or downright evil. The correct "approach" to systems is to live in them, to react in terms of one's experience, and not to try to change them by means of some grandiose scheme or mathematical model. There are all kinds of anti-planners, but the most numerous are those who believe that experience and cleverness are the hallmarks of good management.

Now, the recent interest in the systems approach has concentrated mainly on the scientific version, because this appears to have created some novel ideas and techniques, and the main point of this book will be to discuss these innovations. But since the other three approaches are still active and vocal, we'll let them pose the questions and criticisms. The plot of the debate begins with a conversation between an efficiency advocate and an up-to-date scientist. Then we listen in more detail to what the scientist means by a "system" and how he sometimes can apply his meaning very precisely, e.g., in a mathematical model. He can do this most successfully when the problem is well structured. But the most critical problems of today's systems, war, poverty, racial disturbances, national and state budgets, are all poorly structured. Nevertheless, the scientist believes he can apply the *logic* of his approach to these more poorly structured areas, and we shall watch him try to do so in the areas of budgeting and planning. Finally, we shall see that the scientist runs into the greatest difficulty in trying to cope with human values, and especially the conflicts of values. He may try to resolve his difficulties either by an extension of economic considerations (monetary values), or by behavioral science. Here is where the opposition to his approach by the humanist and the anti-planner becomes strongest. First, then, we turn to the efficiency approach to see why the scientist thinks it's "old-fashioned."

2. EFFICIENCY

You may have suspected that the example of the systems approach given in the last chapter was a bit sneaky. There I described how one kind of *scientific* approach, used by the engineers in charge of designing a rocket to land an object on the moon, was able to cordinate all of the components of the system so that the basic objective of the system could be accomplished with a minimum of delay. In some sense this example was a bit sneaky because the doubters can question whether the objective itself is worthwhile. Indeed, if we look at this objective in the context of worldwide problems, then it's certainly sensible to ask whether the true worldwide objectives are best served by our sending an expedition to the moon.

The objection is well taken, and the systems scientist will need to say some things about the meaningfulness of the objective of the system before we can appreciate his approach. In his own terms, if we unquestioningly accept an objective as a valid one, then we may waste many hours on details that are in the main terribly irrelevant.

But there seems to be one overriding objective of all managers of systems, namely, the efficiency of operations; or in other words, the objective of reducing costs. Any manager who is alert looks around his system and discerns where unreasonable wastes are occurring; if he's a good manager, he does his best to eliminate these wastes in order to reduce the total costs of operation of the system. As Taylor [1] and his "scientific managers" saw it, there is an efficient way to do a job, and it's up to the manager and his staff to find it.

The "efficiency expert" may simply be the housewife who figures how to run the household within the family budget, or he may be the consultant of a large industrial firm or a government agency who determines how to keep the cost within the budget.

Cost means the using up of resources. It is usually measured in terms of dollars, but very often the real costs can be thought of in terms of time or physical resources or men. Every time a dollar is spent, or a man is used to perform a task, or a physical resource is burned up in some way, then there is a lost opportunity for doing other kinds of jobs. When the manager operates under a budget he has to think that every dollar spent means a certain segment of the total budget is used up and is lost forever. Consequently, he is concerned to keep the efficiency of his system at the highest peak so that every dollar spent is spent correctly and contributes to the real objectives of the system. And note that this cost-minimization objective holds for every system; systems, says the efficiency enthusiast, must be run efficiently or they're not worth running at all.

The philosophy of the efficiency approach to systems is based on the idea of "the one best way," i.e., the correct way to perform a task. If the task is the manufacture of a product, then the efficiency approach consists of timing every motion and designing the steps of the task so as to minimize the time. The

[1] Frederick W. Taylor, a pioneer in the field of job analysis, time-and-motion study, etc.

result is a "tight ship" that performs in the best possible way. In most cases, of course, the "one best way" is not known, but, says the efficiency enthusiast, every manager should do his best to approximate it.

Much has been written and debated about the efficiency approach, especially by engineers and humanists. It is an approach that leads naturally to automation, because in many instances, if we know the one best way, we also know how to design a machine to do the task. The humanist is horrified at the resulting degradation of human dignity and the neglect of deeper human values. He points out that large "gains" in efficiency usually lead to unemployment or else plain drudgery.

But we have heard so much from the humanist about the evils of efficiency that a different opposition to the efficiency philosophy may be more enlightening. This is the opposition of the science approach to systems. The argument is that concentration on efficiency *per se* may be a very inefficient way to manage a system, *from the overall point of view*. In other words, the "one best way" may not be the optimal way for the whole system. This opposition to "scientific management" sometimes uses a very similar name, "management science"; but the two philosophies are poles apart.

To see how the opposition formulates its argument, consider an efficiency expert in an industrial firm who, after wandering through the factory warehouse, discovers stacks and stacks of unused inventories sitting there day after day. To him, the inventories mean tied-up money. If the inventories cannot be sold, then it's much as though the manager had to take dollars out of the bank and pile them up in a warehouse, thus letting the dollars sit there without earning any income whatsoever.

Or the expert may be struck by the fact that on occasion many of the work force are idle. This situation becomes apparent when he looks over a large office and sees how many of the secretaries and clerks are engaged in useless tasks or else are not doing anything at all.

Perhaps among the most startling and obvious wastes he may

notice are the countless pieces of idle equipment. Anyone who has passed a railroad switchyard may be struck by the number of freight cars sitting idly on the tracks patiently waiting for an engine.

A more subtle type of cost wastage occurs in the operations of hospitals and social-welfare agencies. To the efficiency-minded, it seems that the hospital staff, while busy during one part of the day, is forced to be unnecessarily idle at other times, when the emergency cases are not so frequent. In welfare agencies one discovers that some of the cases being served are not cases of real need at all. The instances of dishonesty in appealing for welfare assistance make the efficient manager think of the enormous waste of money in doling out public funds to families who can easily care for themselves.

Faced with all of this evidence of waste, the efficiency expert is prepared to look around for ways in which vital inventories can be sold and removed even at reduced costs. When he sees idle men standing around, his inclination is to institute labor-force reductions. When equipment sits idle for days at a time, he wants to sell it or to share it with other agencies or companies. When office procedures seem mainly to include meaningless bureaucratic tasks, the manager wants to cut the office staff. When illegitimate welfare cases come to his attention, his inclination is to cut back on welfare budgets. When obvious inefficiencies occur in the operation of hospitals, his inclination is to decrease staff and increase work efficiency.

Now, the efficiency expert is "right." The operations of every industrial firm or government agency are inefficient, and it is always possible to increase efficiency by revised work methods or the reduction of the work force. The amount of "slack" in our government and industrial organizations has never been estimated, but that it is very large no one questions. Hence "cost-reduction programs" always make sense *within the narrow confines of each division of the organization.* This is why politicians can always make hay on a kind of seasonal cycle of "efficiency programs": in the "summer season" there are plenty of funds, but in the "winter season" the funds are frozen.

Now, the management scientist's argument against "efficiency" is that it is always conceived in relation to a small segment of the social organization. Mere attention to cost reduction by itself, he says, may do the very opposite of what the manager intends. In fact, cost reduction in many instances may actually *increase* the system's total cost.

Suppose we look at some clearcut examples in which a strict cost-reduction policy leads to an increase in the *total* cost of the system. These examples, says the management scientist, show how efficiency considerations by themselves stifle thinking about total system performance. In fact, cost reduction has very much the flavor of the listing of troubles and problems that we discussed in the first chapter. If you sit down and list all of the problems that bother you, you may find that the exercise is a waste of time because you haven't been thinking about the basic objectives of your life and of the organizations to which you belong. In the same way, if you begin concentrating your attention on all of the things that give rise to expenditures of various kinds, you are wasting your time thinking about only one aspect of the total operation. As a result, you will find yourself going down some very bad pathways.

As a first example, consider an airport where the planes land or take off on a single airstrip during the day. To make the example very simple at the outset, suppose that the planes arrive or take off exactly one minute apart, and suppose that it takes exactly one minute for a plane to clear the airstrip. The efficiency expert might be very proud indeed of the operation of the airport. He could see that the airstrip is in continuous use and yet there are no awkward conditions occurring in terms of the stacking of aircraft. As each plane arrives or takes off, it occupies the airstrip for a fixed amount of time, and gets off it, just in time for the next plane to land or take off.

But now suppose the situation is changed a little bit. Let's suppose in our example that the planes arrive or take off *on the average* once every minute and *on the average* they take one minute to clear the airstrip. The phrase "on the average" means that on occasion two or more planes will come very close to each other and this will be balanced by occasions when the arrival or take off occurs at some distance apart in time. If one were to clock the arrivals or departures, one would find that the average still is one per minute, but that in a number of instances there are several planes requiring service at the same time, whereas in other instances no arrivals or departures occur for say two or three minutes. The same situation applies to the use of the airstrip. On some occasions the planes are slower and occupy the airstrip for a longer period than one minute, whereas on other occasions the pilot and the wind conditions permit the plane to clear the airstrip faster than the average.

What would happen in this case? The results are rather startling and can be worked out by means of what the management scientist calls a "probability model." The probability model operates on much the same principle as a slot machine; it tells us the probability that certain events will occur. What interests us in this example are two specific kinds of event: the idle airstrip and the waiting plane. If the airstrip is idle too much of the time, it is "inefficient"; but if the planes have to wait too long, then their performance will also be inefficient. One inefficiency has to be balanced with the other, and *this* is the point that the management scientist believes the efficiency expert misses. In the case of the airport it can be shown that, if the variation in service requirements and occupancy of the airstrip follows the usual pattern, *the waiting line of aircraft will eventually increase without limit.* In other words, the system becomes more and more inefficient in terms of waiting time, even though the airstrip is used "efficiently." One could not predict this result without the use of probability theory, but the result is true nonetheless and, says the management scientist, should be a warning to the overefficient manager.

Thus, if the management scientist suggests to the efficient manager of this airport that he install an additional airstrip to take care of the waiting planes, the manager who is overly concerned with cost reduction will resist the suggestion. He will point out that at certain times of the day the airstrip is not in use at all. In other words, here is a "wasted" piece of equipment that lies idle for periods, and yet someone has the nerve to suggest adding more equipment. But the manager is concentrating on only one aspect of his total system. If he begins to think about the whole system, he will see that the uncertainties concerned with arrivals and with servicing make it absolutely essential that at some time during the day the airstrip be idle. This "inefficient" idleness is absolutely essential so long as we cannot exactly control plane arrival and service times. Now, it might of course be possible to enlarge the system so that the planes could be scheduled to arrive and depart exactly one minute apart but the costs of doing this might be far greater than the costs of simply adding one more service unit.

The management scientist can construct a very simple but illuminating table to illustrate the situation for the manager of the airport. This table shows the average amount of time that planes would have to wait, given that there were one, two, or three airstrips available, under the condition that the average demand for service is one per minute and the average time to clear the strip is one minute. In developing these statistics the management scientist is urging the manager to think of waiting planes as a cost. If the manager broadens his perspective to include the costs of waiting as well as the idle service units, then he may feel perfectly justified in installing additional airstrips on an "efficiency" basis.

To the proponent of efficiency this example will seem to labor the obvious, kick a dead horse, or otherwise abuse one's patience. "Clearly," he will say, "no one in his right mind intends to become so efficient that necessary service is neglected. I was all along urging an increase in efficiency *with the same level of service*. That is, I was saying that we can always remove the slack but still perform all the necessary tasks. In your example, the airport manager is ridiculous. Of course, a new airstrip is essential here, because there is no other way to keep the service at the desired level." To this rebuttal the management scientist replies that the efficiency expert has begged the entire question, namely, *what* level of service is desirable? Obviously, some planes will have to wait, or some airstrips will have to be idle, or both. What combination of waiting and idleness is optimal in the *whole* system? To label either idleness or waiting as an "inefficiency" is to miss the central idea of system planning for the scientist; neither is "inefficient" by itself. The *total* cost of the airport's operation has to be calculated for each policy. And this, says the management scientist, can be done only by means of a model of the system.

The management scientist's precise model for thinking about the costs of service units seems first to have been done by the telephone companies when they began to consider the amount of service that they should supply to customers in the central telephone offices. When a subscriber picks up the telephone and waits for an operator to respond, he is very much like a plane approaching an airport. Suppose the managers of the telephone companies were concerned about idle operators, so that when they went through the telephone office and saw many operators waiting for calls to come in, they felt inclined to cut costs by reducing the number of operators. If they tried to match the average number of calls with the average time to serve a customer, then they'd run into exactly the same situation that occurred in the airport example. The number of waiting customers would begin to increase and the waiting time of the customer for the response of an operator would also increase. In terms of *overall* cost this implies that a cost-reduction program might increase the total cost. The problem the telephone system faces is to "optimize" the total effectiveness of the system, i.e., minimize the cost of waiting plus the cost of idle time. This can be done only by means of a "waiting line" model of the type described above.

The management scientist's point therefore is a very telling one. From the mere fact that there are idle men or idle equipment one cannot infer that the system is operating inefficiently from the *total* cost point of view.

The same idea can be illustrated in many other ways. There is a wonderful story that's told in systems-science circles about two managers of a large company who took a course in operations research. During the course they were told about a mathematical technique for studying transportation problems. The instructor explained to his management students that, if they had various supplies at several factories which were to be delivered to a set of warehouses or retail outlets, then there was an explicit and precise technique that would tell them how to minimize the cost of transporting material from the factories to the warehouses. This technique specifies exactly how much should be sent from a given factory to a given warehouse in order to minimize the cost of transportation. To apply the technique, one has only to estimate the cost of transportation from each factory to each warehouse. The result is a maximum of efficiency in transportation.

When the managers arrived home, they were so inspired by their course that they asked one of their captive mathematicians to work on the problem. The firm gathered together the necessary cost information and the mathematician put the pieces together into a mathematical model and politely requested a computer to grind out a solution. Much to the managers' disappointment, the new mathematical technique saved only 50,000 dollars a year of the total transportation cost. The managers had expected a much larger saving, because, although mathematicians still come fairly cheap, computers do not, and the total cost of computation was larger than the so-called savings.

These were very efficiency-minded managers, but they were concerned about only one aspect of their operations—namely, how to reduce the costs of transportation. Since computers can't be wrong, it must have been the mathematician who erred. Hence, the managers appealed to another research team to go over the results to detect their mathematician's mistakes. The research team was quite willing to do this, at a fee; they checked the mathematician's calculations and reported essentially the same saving. But while the managers were waiting for the results the research team began to delve a little more deeply into the managers' problems. They began to ask about the policy of production in each of the factories and the problems of transportation of all materials into the factories. They also began asking very astute questions about why a certain warehouse needed certain materials at all. In other words, they began to broaden their view of the system and to argue that the total system consisted of materials which went into the factories, from the factories into warehouses, from the warehouses to various retail outlets, and thence to customers. When this total picture was put together, it became apparent that the current policies governing the amount stored in each warehouse were irrational in terms of the total operation: certain warehouses should not receive the items that they traditionally received, whereas others should. In effect, then, the attempt to make the transportation subsystem "efficient" was an attempt to do "precisely" the wrong thing.

When all of the pieces of the system were put together, it was possible to generate a cost saving for the total system in the tens of millions, and most interesting of all, it became apparent, as the total system was examined, that the transportation costs from factories to warehouses should be increased.

The same story can be told over and over again in many organizations. It's true there is nothing more annoying than seeing large inventories lying around idle. This has led many an industrial manager or logistics expert to suggest across-the-board reductions in inventory. One way to accomplish this is to introduce special sales or to sell the inventory at salvage prices. But this problem of inventories is very much like that of the airport. Idle inventories simply have to be held to take care of unpredictable demands. We can't always predict exactly the arrival of airplanes, and we can't always predict exactly the demands on most inventories. If, for example, the inventories are held to supply maintenance crews, then a shortage in inventory may mean tying up a large piece of equipment for many days because the parts are not available. It simply is not "efficient" operation from the *total* systems point of view to get rid of idle supplies in order to release some cash for some other purpose. For the management scientist, the problem of holding inventories is a problem involving both the cost of inventory and the cost of shortage, and this problem can be solved only by a model. If the service objective is forgotten in an effort to reduce costs, then the total system cost may increase even though a cost reduction has occurred.

The spirit of the efficiency approach, nonetheless, does not die. Across-the-board cost reduction still seems to appear, much like a haunting ghost, in many political situations. For example, a governor of California used his "iron hand" to institute ten-percent cost reductions across the board in state offices and universities. The U.S. Congress has recently been putting pressure on the President to cut costs. If cost reductions are accomplished by releasing "idle" personnel or "idle" inventories or equipment, then the governor and the Congress are apt to find that the cost reductions become very expensive indeed. But the reason the efficiency approach will survive is that it is based on the truism that most management is careless or inept and can be improved.

For the management scientist there is a "lie" that lies hidden in so-called cost data of the efficiency expert. If the cost data have been collected in terms of direct expenditures and the manager tries to reduce these expenditures, he will find that what he considers to be real costs are illusory. Without a measure of the total system performance against which he can compare costs, his cost data mean nothing whatsoever. And if he tries to reduce his "costs," he'll find that his performance goes down.

This does not mean of course that all idle inventories, idle men, and idle equipment are beneficial to the system. All it means is that idleness by itself is not the only thing to consider when one is thinking about how a system is performing. Idleness by itself is not even a symptom of trouble. Unless one has a broader context in which to think about the system, then it's useless to think about idle inventories and men and equipment.

According to the management scientist, the costs that a manager incurs are always opportunity costs in the sense that, when he uses some dollars or men or equipment for a given purpose, he is sacrificing the use of these dollars, men, and equipment for some other purpose. The real purpose must be gauged in these terms. If equipment is idle and he wishes to put it to use in some other context, there is a cost of doing this. If it is used for another purpose, and hence the equipment no longer can be used for its original purpose, then it's this loss of opportunity which must be thought about as we begin to think about costs.

The proponent of efficiency will still want to have his say. He will point out that all this talk about the total system is largely idealistic. When there is a mess, we should clean it up. If a house catches on fire, it's foolish to spend time thinking about alternative uses of equipment. The right prescription is: Put out the fire and save lives. Wherever inefficiency, waste, lawless aggression and the like occur, make effort to remove them.

The management scientist doesn't disagree with this viewpoint, so long as it's clear—in terms of the total system—what is really inefficiency or danger, or what is not.

You will begin to see that the management scientist's thinking about systems is no easy matter. According to him, we are always obliged to think about the larger system. If we fail to do this, then our thinking becomes fallacious.

But thus far the management scientist has been playing the role of a critic of the philosophy of efficiency. If we're to understand his approach to systems, we'll have to learn what he thinks a system is and hence the steps we should take in thinking about it. He may not be able to come up with a thoroughly satisfactory model for all systems, but at least he may help to guide our thinking and keep us out of narrowly defined pathways.

3. SYSTEMS

There is a story often told in logic texts about a group of blind men who are assigned the task of describing an elephant. Because each blind man was located at a different part of the body, a horrendous argument arose in which each claimed to have a complete understanding of the total elephantine system.

What is interesting about this story is not so much the fate of the blind men but the magnificent role that the teller had given himself—namely, the ability to see the whole elephant and consequently observe the ridiculous behavior of the blind systems describers. The story is in fact a piece of arrogance. It assumes that a very logically astute wise man can always get on top of a situation, so to speak, and look at the foolishness of people who are incapbale of seeing the whole. This piece of arrogance is what I called "management science" in the last chapter.

The arrogance cannot be allowed to remain unchallenged. Only if we could be sure that the objectives of the management scientist were pure and really in line with those of the total system, and only if we could be sure that he had the observational powers comparable to those of the observer of the blind men could we feel that the scientist had the ability to see the whole.

But in the spirit of the debate, let's allow the management scientist to describe how he climbs to the vantage point from which he can view the whole system. His method is one of defining carefully what he's talking about. He begins with the term "system." Although, he says, the word "system" has been defined in many ways, all definers will agree that a system is a set of parts coordinated to accomplish a set of goals. An animal, for example, is a system, a marvelously contrived one, with many different parts which contribute in various ways to the sustaining of its life, to its reproductive pattern, and to its play.

In order to make this definition more precise and also more useful, we have to say what we mean by "parts" and their coordination. Specifically, the management scientist's aim is to spell out in detail what the whole system is, the environment in which it lives, what its objective is, and how this is supported by the activities of the parts.

To develop this thinking further, we shall have to lay out a series of thinking steps, much as any manual of logic or rhetoric attempts to do. The reader should bear in mind, however, that these steps are by no means steps that must be taken in sequence. Rather, as one proceeds in thinking about the system, in all likelihood it will be necessary to reexamine the thoughts one has already had in some previous steps. Logic is essentially a process of checking and rechecking one's reasoning.

With this in mind, we can outline five basic considerations that the scientist believes must be kept in mind when thinking about the meaning of a system:

1. the total system objectives and, more specifically, the performance measures of the whole system;

2. the system's environment: the fixed constraints;

3. the resources of the system;

4. the components of the system, their activities, goals and measures of performance;

5. the management of the system.

It goes without saying that there are other ways of thinking about systems, but this list is both minimal and informative.

The objectives of the overall system are a logical place to begin, because, as we have seen, so many mistakes may be made in subsequent thinking about the system once one has ignored the true objectives of the whole.

At the outset, however, we must beware of a confusion about the word "objective." The inhabitants of systems dearly love to state what their objectives are, and the statements they issue have a number of purposes that are quite independent of the performance of the system. The president of a university wishes to attain as large a budget as possible for the university's operations. As a consequence, he must appear before a number of legislative committees and before the public, and in these appearances he must state the objectives of the university in as attractive a manner as possible. His aim is to attain as much prestige and as much political power as he can in order to obtain for his university the largest possible budget for its operation. Hence, he speaks of quality of education, eminence of faculty, public service, and the like. Similarly, the head of a large business firm in his public appearances must present a glowing picture of the objectives of his firm. He does this not only to attract customers but also to attract satisfactory investment funds.

In many firms and government agencies, these vague statements are often called *the* objectives, but from the scientist's point of view they are obviously too vague and also somewhat misleading. For example, if we take the public statements too seriously, we may be misled in identifying the real as compared with the stated objectives of the system. The president of a university is apt to make us think that the sole objective of the university is the creation of new knowledge and the teaching of knowledge to eligible students. The head of the business firm is apt to make us think the sole objective of his firm is the maximization of net profit subject to considerations of public service.

Now the scientist's test of the objective of a system is the determination of whether the system will knowingly sacrifice other goals in order to attain the objective. If a person says that his real objective in life is public service and yet occasionally he seems quite willing to spend time in private service in order to maximize his income, then the scientist would say that his *stated* objective is not his *real* objective. He has been willing to sacrifice his stated objective at some time in order to attain some other goal.

A common fallacy in stating objectives is to emphasize the obvious. For example, consider a medical laboratory that tests specimens which doctors send in. What is the objective of the laboratory? One obvious answer is to say that the objective is to make as accurate a test as possible. But the real objective is not "accuracy" but what accuracy is good for: improving the doctor's diagnosis. Once we look ahead to the desired, concrete outcome, says the scientist, then we can ask ourselves how important the objective really is. In some cases, improved accuracy may not be worth the cost, i.e., the sacrifice of other objectives.

Of course, it is no easy matter to determine the real objectives of a system, any more than it is an easy matter to determine the real objectives of an individual person. We all hide our real objectives because in some cases they are hardly satisfactory ones from the point of view of other people; if they are widely publicized they may be harmful in terms of our prospects of attaining various kinds of support in our lives.

In order to clarify the matter, the scientist needs to move from the vague statement of objectives to some precise and specific measures of performance of the overall system. The measure of performance of the system is a score, so to speak, that tells us how well the system is doing. The higher the score, the better the performance. A student in class often comes to think of his objective as the attaining of as high a grade as possible. In this case the measure of performance becomes quite clear, and it's interesting to many a teacher to note that students seek to attain a high grade even at the sacrifice of a real understanding of the content of the course. They seek the high grade because they believe that high grades will lead to scholarships and other opportunities in the future. Their *stated* purpose is to learn, but their *real* measure of performance is the grade.

In the same manner, if we look very carefully at certain cities we may come to expect that the real objective of the government of the city is to sustain the opportunities of the high-income citizens by providing them with satisfactory areas for living and satisfactory resources and space for their work. Thus the claims that the city is trying to serve *all* the citizens are refuted by the city supervisor's willingness to sacrifice these aims in favor of sustaining the opportunities of the higher-income bracket. The *real* measure of performance, then, is the city's capability of keeping large industries within the city boundaries and keeping the level of income of the high-income group as high as possible.

Similarly, in the case of certain firms, some economists believe that the objective of the firm is not net profit but growth of personnel or gross profit, these two measures representing the size of the empire, so to speak. The point is that, in these firms, the managers are willing to sacrifice a certain amount of net profit in order to increase the size of the firm, in terms of either personnel or gross earnings or assets.

It will come as no surprise that a careful study of certain colleges and universities indicates that the true measure of their performance is not in terms of education but in terms of the number of students graduated.

These remarks give us some clue about the character of the management scientist. He wishes to strip away all the folderol nonsense about "my heart is pure and I'm out to serve mankind, or motherhood." He wants to see what this beast called the system is really up to, and he can do so only by carefully watching what it actually does, not what it says it does. Furthermore, he thinks he can strip away enough of the noise of confusion and uncertainty to see a central "measure" or "score" for the system.

We can already begin to hear a rumble of complaint from his

opponents. Some of them will want to point out that an additional distinction must be made between the *real* objectives and the *legitimate* objectives of the system. The legitimate objectives of the system have to do with the morality of the system objectives. For example, the management scientist may define the objective of a highway system in terms of what he calls "thruput," meaning by this the number of cars that are able to pass over specified segments of the highway within a given period of time. However, the objective itself may not be "legitimate" from a social point of view, not only because of the cost of accidents but also because of the inconvenience that may occur when cars pour off the exits of the freeways, and the ugliness of the freeway system itself.

But to the thorough-thinking management scientist, this objection is not a serious one. In thinking about systems, he replies, we must move from what is often the real objective of the system managers to wider considerations. We may in fact have to begin to consider how to put the cost of accidents and ugliness into our measures. Intangible as these may be, he says, we shall see that the measurement of them is really not so difficult as might appear at first sight. In fact, there are some excellently worked-out cases in which highway engineers, as well as the designers of aircraft, have developed measures of the cost of an accident, in terms of the lost capability of the individual in earning income throughout the rest of his life. To the humanist, this may seem a very crass way of putting a number on the loss of a limb or a head, but to the management scientist it is the only practical way in which we can think about the so-called intangible aspects of the systems. In other words, he says, if we want to *think* about how loss of life or happiness or beauty is related to system performance, we can't simply say that these are so elusive that they cannot be defined, because by saying this we mean that we don't want to think about them at all. In order to think about them satisfactorily, we are going to have to be explicit and make our stand on the way in which these aspects of systems enter into measures of system performance.

The management scientist is turning out to be persistent, at least, though his persistence may leave a number of his listeners uncomfortable. He is not only persistent but also alert. With experience, he becomes convinced that the "obvious" measures of performance are not the real ones.

One example of the fallacy of the obvious has quite an ironic twist. In the field of health, with the advent of vaccines for many "classic" diseases, it seems obvious that the "health system" should eliminate the blights. Recently, for example, steps have been taken to eliminate measles. It looks as though the measure of performance should be the reduction in percentage of children who come down with measles, possibly weighted by the reduction in the seriousness of the cases. A newspaper editorial points out that measles in the Near and Far East often proves fatal; consequently, goes the argument, a "success" of the system according to the above measure will result in reduction in infant mortality, and consequently will produce an "intolerable" increase in population in underdeveloped areas. Here again the character of the "whole systems" thinker becomes apparent: perhaps it is "better" to let measles do its ugly work than to allow the starvation resulting from the population explosion. This is just another example in which costs must be included in the measure of performance.

Thus in the determination of a measure of performance, the scientist will seek to find as many relevant consequences of the system activities as he can. Admittedly, he too will make mistakes and will have to revise his opinion in the light of further evidence. But his persistence and alertness, and his intent to be as objective as possible, will enable him, he believes, to minimize his errors.

Supposing that some success has been attained in determining the system objective ("measure of performance"), the next aspect of the system the management scientist considers is its environment. The environment of the system is what lies "outside" of the system. This also is no easy matter to determine. When we look at an automobile we can make a first stab at estimating what's inside the automobile and what's outside of it. We feel like saying that what lies beyond the paint job is in the environment of the automobile. But is this correct? Is it correct to say, for example, that what lies beyond the paint job of a factory is necessarily outside of the factory as a system? The factory may have agents in all parts of the country who are purchasing raw materials or selling its products. These are surely "part" of the total system of the factory, and yet they are not usually within its walls. In a more subtle case, the managers of the factory may belong to various political organizations through which they are capable of exerting various kinds of political pressures. Their political activities in this case certainly "belong" to the system, although again they hardly take place within the "shell" of the system. And, returning to the automobile and considering what it is used for, we can doubt whether its paint is the real boundary of its system.

Perhaps, after all, the super-observer of the blind men trying to describe the elephant was himself rather blind. Does the skin of the elephant really represent the dividing line between the elephant and its environment? Maybe an understanding of the habitat of the elephant is essential, and perhaps the habitat should be regarded as part of the elephantine system.

Marshall McLuhan has pointed out that in the age of electric technology the telephone has actually become a part of the individual person. Indeed in many cases, it would be hard to differentiate between the ear and the telephone that serves the ear. His point is that we cannot "cut off" the telephone any more than we can cut a person's ear off in any satisfactory way. The telephone is part of the system that we call the individual person.

Hence the scientist has to have a way of thinking about the environment of a system that is richer and more subtle than a mere looking for boundaries. He does this by noting that, when we say that something lies "outside" the system, we mean that the system can do relatively little about its characteristics or its behavior. Environment, in effect, makes up the things and people that are "fixed" or "given," from the system's point of view. For example, if a system operates under a fixed budget that is given to it by some higher agency and the budget cannot be changed by any activities of the system, then we would have to say that the budgetary constraints are in the environment of the system. But if by some organizational change the system could influence the budget, then some of the budgetary process would belong inside the system.

Not only is the environment something that is outside the system's control, but it is also something that determines in part how the system performs. Thus, if the system is operating in a very cold climate so that its equipment must be designed to withstand various kinds of severe temperature change, then we would say that temperature changes are in the environment, because these dictate the given possibilities of the system performance and yet the system can do nothing about the temperature changes.

One of the most important aspects of the environment of the system is the "requirement schedule." In the case of an industrial firm this consists of the sales demands. Of course in some sense the firm can do something about the demands by means of advertising, pricing, and the like. But to the extent that the demand for the firm's products is, so to speak, determined by individual people outside who are the customers of the firm, then the demand lies in the environment of the system, because it is a "given" and because its nature influences system performance.

Here again we get some insight into the character of the management scientist. The environment is not the air we breathe, or the social group we belong to, or the house we live in, no matter how much these may seem to be outside us. In each case, we must ask, "Can I do anything about it?" and "Does it matter relative to my objectives?" If the answer to the first question is "No" but to the second is "Yes," then "it" is in the environment.

The management scientist is normally a very careful person, and he knows how difficult it is to determine the system's environment and that the problem needs to be reviewed systematically and continuously. Often systems fail to perform properly simply because their managers have come to believe that some aspect of the world is outside the system and not subject to any control. I was recently watching a television show whose theme was that the poor pay more than the rich for home products. The purpose of the show was to indicate how stores increase prices in poor neighborhoods, and specifically how credit agencies often require the poor to pay far higher interest rates than do the rich. In its thinking about how to overcome this community difficulty, the program urged an education of the poor, so that they would not be duped by salesmen of freezers, television sets, and the like. In its analysis of how it comes to happen that the credit system is so unfair to the lower-income groups, the program described how the credit system is controlled by banks and ultimately by policy makers on Wall Street. But the program designers did not even think it advisable to educate any of the banks and Wall Street with respect to the impact of their policy on the poor communities of cities. In other words, the program designers had taken the policies of the banks and of Wall Street to be in the environment of the credit system, and hence not subject to any change. From the management scientist's point of view, it's clear that some mistake may have been made here. It might in fact be possible, if one were to employ a systems approach to credit policies, to show how the rather rigid policies with respect to low-income groups generate a series of community problems which themselves badly affect the operation of the community and hence increase the costs of operation of large industries and even of the banks themselves.

Next we turn to a consideration of the resources of the system. These are *inside* the system. They are the means that the system uses to do its jobs. Typically when we turn to the measurement of resources we do so in terms of money, of man hours, and of equipment. Resources, as opposed to the environment, are the things the system can change and use to its own advantage. The system can decide which of its men shall work on which jobs, or how its money shall be spent on various activities, or what the time limits will be on various kinds of activity.

Just as it is difficult to think adequately about the environment of the system, it is also quite difficult to think adequately about its real resources. I have already had something to say about this in the illustration of idle time, idle equipment and idle men. Here the manager, overanxious about his resources, may come to believe that idle men and idle equipment imply an unused resource, and if he sets about too energetically to change the idleness into work, he may find that he is actually decreasing his resources.

Within many systems a very careful analysis is made of resources. The traditional company balance sheet in effect is a listing of the various kinds of resources that a firm has available, especially when these resources can be translated into money: buildings, equipment, accounts receivable, cash, etc. But the management scientist has concluded that the traditional balance sheet leaves out many of the important resources of a firm. It does not give a detailed account of the type of personnel that the firm has available in terms of their educational background and personal capabilities, for example. Something like "good will," which is surely a resource, is often represented by a fictitious number on the balance sheet.

But there is an even more serious objection to the income statement of an organization; this statement is supposed to show how resources were used. The management scientist is chiefly interested in learning from experience, since this is always the hallmark of excellence in science. But the typical income statement hides almost all the relevant information that should be collected if one is to learn from an organization's past. The real lessons to be learned are the lessons of lost opportunities, the possibilities that were never actualized because the resources were used elsewhere. These lost opportunities are the cases that should be watched, but they are practically never described in the operating statement of business organizations.

For the management scientist, the systems approach entails

the construction of "management information systems" that will record the relevant information for decision-making purposes and specifically will tell the richest story about the use of resources, including lost opportunities. Later on we shall look into the design of such a management-information system in some detail.

There is another aspect of resource determination that is quite important in an age of expanding technology: it is quite essential that firms and government agencies pay special attention to technological advances that may be able to increase their resources enormously. We shall have some things to say, for example, about increasing computer capabilities and how these lead in effect to a "free" increase in a firm's resources. In looking and thinking about a system, the management scientist pays attention not only to existing resources but also to the manner in which resources can be increased, that is, to the manner in which the systems resources can be used to create better resources in the future, by means of research and development in the case of hardware types of equipment, or by training and education of personnel, and by various kinds of political activities which will increase the budget and investment potential. In fact, for many systems a component that deals with the increase of resources may be the most important component of the system.

Resources are the general reservoir out of which the specific actions of the system can be shaped. The specific actions are taken by the components, or parts, or subsystems (all these terms being used interchangeably in management science). Components is the fourth item in the "thinking" list on page 30. Here again, says our scientist, our thinking is apt to be blurred by tradition. Organizations are often divided into departments, divisions, offices, and groups of men, but careful examination shows that these are not the real *components* of the system even though they carry labels that seem to indicate that they are. For example, in industrial firms a department may be labeled "production"; this should lead us to think that only within this component can one find the manufacture of products. Another department will be labeled "marketing"; one should therefore believe that only in that department would one find the activities dealing with distribution and sales of products. And yet in many firms the distribution function must be conceived as part of the production component simply because it would be quite impossible to determine how the distribution of products should occur independent of the way in which the products are made. And perhaps the production department has a great deal to do with the manner in which products are sold simply because production must deal in many cases directly with the customer in satisfying his orders. If the customer is badly disappointed, then the activities of the production department may decrease sales.

It is for this reason that in thinking about systems the management scientist ignores the traditional lines of division and turns instead to the basic "missions" or "jobs" or "activities," all of these labels being used to describe the same kind of thing, namely, the rational breakdown of the tasks the system must perform. Thus in the case of a city or a state, the basic missions may be defined in terms of health, education, recreation, and the like. If they are so defined, the scientist sees that many different agencies are engaged in the mission of health, even though their labels may not so indicate. For example, the motor-vehicles department of a state may have a good deal to say about the steps that should be taken to identify an individual on the highway who is intoxicated or is overcome by a stroke. The scientist wants to say, therefore, that the motor-vehicles department is actively engaged in the health mission. In the same fashion the educational function of the state takes place not only within the department of education, but in many other departments which are engaged in various kinds of training programs for their own personnel and educational programs for the public by means of brochures, short courses, TV demonstrations, and the like. The overall valuation of the education mission therefore cannot take place within traditional department lines.

There is bound to be a lot of resistance to this mission-oriented view of the "components" of a system. In terms of politics, the head of a department knows that his department is a unit and a distinguishable part of the total organization. He has to do battle for budget and personnel with other "components," and he is judged in terms of how well his "part" has done in supporting the total organization's goals. Furthermore, the people who work in his department identify themselves with the department, not the so-called mission, which merely exists in the head of the management scientist. This is especially the case in universities. It may be that mathematics and philosophy are widely studied and practiced in all the fields of learning, but *the* departments of mathematics and philosophy define what these subjects "really" mean—i.e., really mean to the true mathematician and philosopher.

The management scientist, however, is not a very sympathetic fellow. He can see that political and personal ambitions influence people into believing that the parts of the system should be as independent as possible. People want to say that "education" should take place in a quite separate department from "health" or "recreation." But the management scientist believes that this is a fallacious way to think about the matter. Normally the educational activity does have a good deal to do with health, and health has a good deal to do with education. The proponents of a clear separation of function may urge, therefore, that we think about other kinds of functions which are more separable and in which separate measures of performance can be generated, thereby preserving the integrity of the department. This idea is frequently carried out in machine design in which each component of a machine has a specific function to perform and the performance of a given part is as independent as possible of the performance of other parts. Even in machine design, however, this may not be a feasible way to approach the problem.

Why is the management scientist so persistent in talking about missions rather than departments? Simply because by analyzing missions he can estimate the worth of an activity for the total system, whereas there is no feasible way of estimating the worth of a department's performance. He needs to know whether one activity of a system component is better than another. But if a department's activity belongs to several larger missions, it may not be possible to distinguish its real contribution. This is why the management scientist is so skeptical about managerial accounting, in any of its various forms. The managerial accountant wants to generate "scores" of departmental performance, or "cost centers" which can be examined for their utilization of resources. But insufficient thinking goes into the identification of these scores and centers in terms of their real contribution to the total system objective.

But why do we need components at all? The management scientist would like to look at each choice of the whole system in a direct way, without having to subdivide the choice. But this is not feasible. Consequently, the real reason for the separation of the system into components is to provide the analyst with the kind of information he needs in order to tell whether the system is operating properly and what should be done next. As we shall see, the management scientist thinks he has succeeded reasonably well in certain cases in identifying the real components (missions) of a system. Unfortunately, to date, in most city and state governments there is no adequate systems analysis of the total system in terms of real components; for historical reasons the state and city governments are divided into departments and divisions that often have no relevance to the true components of the system. As a consequence, says the scientist, the management of our large government systems of states and cities becomes more and more difficult each year. Because the decision making that governs different missions is not centralized, the real missions of the state, e.g., in terms of health, education, recreation, sanitation, and so on, cannot be carried out because there is no management of these missions. One of the greatest dangers in component design is the rigidity which has occurred so frequently in the political designs of the cities and states. The assignment of responsibilities becomes fixed by law and impossible to break. What occurs is a kind of hardening of the communication arteries and the disease that sets in is well known to most administrators. Even the most obvious plans for the various missions of the city and state cannot be carried out simply because there is absolutely no way to break up the rigidities of the system that have occurred because of political history.

It goes without saying that our management scientist is antipolitical, simply because so much of politics thwarts the rationality of his designs. He goes so far as to say that city, state, and federal governments cannot be regarded as "systems," because in their design there is no rational plan of the components of the system and of their operation. Nevertheless, there are notable exceptions. Some governmental departments, e.g., the Department of Defense and the National Aeronautics and Space Administration, have taken the "system challenge" quite seriously, as have several state governments. In industry, "system thinking" has often infiltrated quite deeply, even though the concepts of the "whole system"—i.e., the whole corporation—are still very difficult to define. The optimistic management scientist looks forward to a "systems era," in which man at last will be able to understand the systems he has created and lives in.

The ultimate aim of component thinking is to discover those components (missions) whose measures of performance are truly related to the measure of performance of the overall system. One obvious desideratum is that as the measure of performance of a component increases (all other things being equal), so should the measure of performance of the total system. Otherwise, the component is not truly contributing to the system performance. For example, in industrial practice if the measure of performance of a component is in terms of its output per unit cost, then it would be essential to show that, as this measure increases, the total performance of the system increases. If, however, drastic cost-reduction methods are imposed on the component that result in decreased quality of its service or product, then it may very well happen that someone has instituted a measure of performance for the component that does not imply an increase in system performance. For example, a production department may institute various kinds of cost-reduction policies resulting in decrease in inventories. Its output per unit of cost may therefore go up, but the performance of the entire firm may go down simply because the cut in inventory leads to unsatisfactory shortages.

As we shall see, this problem of measuring the performance of a component gets to be a very tricky and difficult one as we go deeper into the design of large systems. Although the simple requirement that the measure of performance of the component should go up as the total system performance goes up seems quite obvious, nevertheless it does not follow that a component can simply push its way along its measure of performance and ignore all of the other components of the system. If some other part of the system changes, say because of the technological improvement, then it may become essential to change the measure of performance of the given component. In office procedures, for example, a typical measure of performance of the office is in terms of the number of letters or documents that are typed per man-hour of office staff. But suppose a systems-and-procedures group shows how various kinds of routine letters can be reduced in size while still containing all relevant information. The measure of performance of the office would go up as a result of this activity but would hardly characterize a true increase in performance of the office. Of course, the point here is that the "office" *per se* is not a true component of the system, since in this case the component should include those who study it in order to improve it.

These considerations bring us to the last aspect of the system, its management. The management of a system has to deal with the generation of the plans for the system, i.e., consideration of all of the things we have discussed, the overall goals, the environment, the utilization of resources, and the components. The management sets the component goals, allocates the resources, and controls the system performance.

This description of management, however, creates something of a paradox for the management scientist. After all, it is he who has been scheming and plotting with his models and analyses to determine the goals, environment, resources, and components.

Is he, therefore, the manager; does he intend to "take over" with his computer army?

The truth of the matter is that he doesn't want to. He is not a man of action, but a man of ideas. A man of action takes risks, and if he fails, not only does he get fired but his organization may be ruined; the man of action is willing to risk fortunes besides his own. The management scientist is typically a single risk-taker: if he fails, he doesn't have to bear the responsibility of the whole organization's failure.

Hence, we've found one chink in the scientist's armor: he doesn't really understand how he himself is a component of the system he observes. He likes to think that he can stand apart, like the elephant observer, and merely recommend, but not act. How naïve this must appear to the politician is hard to say, but certainly the politician's appreciation of the situation is the more sophisticated one. "Mere" recommendation is a fantasy; in the management scientist's own terminology, it is doubtful whether the study of a system is a separable mission.

For the moment, we'll forget this embarrassment of the management scientist and talk instead of other ways in which he can aid the managers of systems. Not only does the management of a system generate the plans of the system, but it also must make sure that the plans are being carried out in accordance with its original ideas. If they are not, management must determine why they are not. This activity is often called "control," although modern managers hasten to add that the term "control" does not imply strong coercion on the part of management. Indeed, many control procedures operate by exception, so that the management does not interfere with the operations of a component except when the component gives evidence of too great a deviation from plan. However, control does not only mean the examination of whether plans are being carried out correctly; it also implies an evaluation of the plans and consequently a change of plans. As we shall see, one of the critical aspects of the management of systems is the planning for change of plans, because no one can claim to have set down the correct overall objectives, or a correct definition of the environment, or a fully precise definition of resources, or the ultimate definition of the components. Therefore, the management part of the system must receive information that tells it when its concept of the system is erroneous and must include steps that will provide for a change.

The control function of management can be studied by the scientist. The late Norbert Wiener compared this function of the management of the system to the steersman of a ship. The captain of the ship has the responsibility of making sure that the ship goes to its destination within the prescribed time limit of its schedule. This is one version of the overall objective of the ship. The "environment" of the ship is the set of external conditions the ship must face: the weather, the direction the wind blows, the pattern of the waves, etc. From the captain's point of view, the environment also includes the performance characteristics of machinery and men, since these are "givens" on any voyage. The ship's resources are its men and machinery, as these can be deployed in various ways. The components of the ship are the engine-room mission, the maintenance mission, the galley mission, and so on. The captain of the ship as the manager generates the plans for the ship's operations and makes sure of the implementation of his plans. He institutes various kinds of information systems throughout the ship that inform him where a deviation from plan has occurred, and his task is to determine why the deviation has occurred, to evaluate the performance of the ship, and then finally, if necessary, to change his plan if the information indicates the advisability of doing so. This may be called the "cybernetic loop"[1] of the management function, because it is what the steersman of a ship is supposed to accomplish. A very critical aspect of a cybernetic loop is the determination of how quickly information should be transmitted. Anyone who has tried to steer a rowboat through rough waters will recognize that, if one responds too quickly—or else too slowly—to the pattern of the waves, he is in real trouble. What is required is an information-feedback loop that permits one to react to the pattern of wind and waves in an optimal fashion.

[1] From the Greek word for "steersman."

Wiener and his followers developed a theory of cybernetics which has mainly been applied to the design of machinery. But it is only natural for the management scientist to attempt to apply the theory to the management control of large organizations.

Thus far we have stated the preliminary case for the management scientist's approach to systems, with some critical comments from the sidelines. Does the management scientist's approach work? If "work" means "use," then it does indeed work. Hundreds of large industrial firms in transportation, power, communication, and materials all use management science under such labels as "operations research," "system science," or "system engineering," "systems analysis," etc. In all cases the avowed purpose of these groups is to approach problems in the spirit outlined in this chapter. Similarly, every section of the military establishment uses management scientists in the design of weapon systems, of information systems (e.g., SAGE and SACCS), of logistics systems, etc. Management science is used extensively in the nonmilitary divisions of the federal government, in public health, in education, in the post office, patent office, National Bureau of Standards, etc. Several states and a number of cities are developing management-science capability as an integral part of their government administration.

It would be wrong to say that all these applications of management science proceed with equal competence or even from exactly the same viewpoint. An illustration may help to enrich the flavor of the approach, however; in the illustration, I'll keep the debate going by allowing the critics their say. The critics often view with alarm, or even disgust, what they regard to be the wholesale and uncritical use of "science" in the important problems of today's government. Some of them want the old tried-and-true method of experience to remain. Some want government to dwindle away. Some are afraid of the inhumane attitude of the scientist. Some think the scientist is simply naïve. They deserve their say, and what they have to say can best be said in the context of an actual illustration, to which we now turn.

4. AN ILLUSTRATION

This illustration of the scientist's systems approach is in the area of transportation. I have chosen the example because it so well illustrates the theme of the last chapter, namely, that every system is embedded in a larger system. This, as we shall see, becomes obviously true of transportation systems.

The illustration lies in the maritime industry, which transports cargo by means of the sea and inland waterways. Although the maritime industry is in some sense one of the oldest of transport industries, its technological advance has been very slow. The manner of loading and unloading cargo that we use today is virtually the same technique that was used in the times of the Greeks, or even by the Indians with their canoes. Essentially one shapes a hull into which the cargo is put in such a way that the vessel does not sink, then one transports the cargo, and eventually hauls it over the sides of the vessel onto dry land.

The maritime industry in the United States in recent years has seen many troubles, and its state has become a concern of the United States government. As a consequence the government, through the National Research Council and the National Academy of Sciences, appointed a series of study teams to look into the maritime industry. A part of the difficulty of the industry arises from the fact that the United States is in a poor position to compete with many foreign countries because of its higher wage rates. Consequently, the specific question raised at the outset of the project was whether or not there were some technological means of improving the performance of the industry that would counterbalance higher labor costs.

Experience shows that the major costs of total handling of cargo in the maritime industry occur at the docks in the loading and unloading operation, and not in that sector of the transport between ports. A team of industrial engineers therefore set to work to determine whether there were some innovative ideas that could be used at the dock in the handling of the cargo. If the research team had proceeded along the lines of pure efficiency, then their objective would have been to try to engineer into the loading and unloading operation technological advances that would have made the operation more efficient, i.e., less costly per unit of material handled.

However, the research team soon realized that the operation of loading and unloading vessels is embedded in a larger system and that this larger system is made up of (1) the companies that own the ships, (2) the labor unions, (3) the so-called casual workers (who are not members of the union but are called in whenever there is a work load to warrant hiring them), and (4) the public at large. If "efficiencies" in the handling of cargo could be attained, then these efficiencies would probably result in decreased demand for labor and increased profits for the companies. In the larger system, therefore, one would expect that the so-called efficiencies that the engineers might discover would, if implemented, bring about serious labor-management problems to the detriment not only of the workers and companies but also the public. Consequently, it was felt desirable to study the port as a system rather than merely confine one's attention to increasing the efficiency of cargo handling of each ship.

At the outset the scientist faces one of the most difficult problems in the entire study, namely, who is the decision maker? For the scientist to describe a system, it is essential that the decision maker be identified; otherwise there is no clear way of determining what the objectives and the environment resources of the system are. Recall that the "environment" of a system is taken to be those conditions which are not in the control of the decision maker, whereas "resources" and "components" are partially under his control. Hence the major breakdown of the system depends on identifying the decision maker. Well, then, who does make the decisions about the operations of a port? To some extent, the companies do, in that they are the ones who design the schedules of the vessels and determine in part what is to be loaded and unloaded. On the other hand, the labor unions obviously have some say in how the port is operated, for stevedores who are members of the union will work on certain days and will not work on other days. The public decides some things through their legislative representatives: each port has a port authority which lays down the conditions under which the loading and unloading of vessels may occur.

Yet from the point of view of this study it is none of these—the companies, labor unions, or the public—who is the true decision maker. The study is being made for the federal government and specifically for two agencies of the federal government. The purpose of the study is to lay out some strategies that these agencies can follow in terms of their recommendations either in the form of legislation or advice to other government agencies. Consequently, the decision maker is taken to be the two agencies for which the study was conducted. This conclusion on the part of the scientists is questionable. It might seem from their argument that in any study the management scientist himself should be the decision maker, as all he really controls is his ability to give advice. The scientist's idea, however, is that for the study team the decision maker is the person or group to whom they report, in this case, the government agencies. Whether this is a proper criterion for identifying the decision maker is itself a matter for debate, to which we shall eventually return.

The scientist's candidate for the decision maker has some rather unusual characteristics. For example, the decision maker has no direct control over any of the operations of the port. Instead, he merely controls the kinds of recommendations that can be made. Nevertheless the objective of the decision maker can be made clear enough. He wishes to come up with a recommendation that has a high probability of being accepted, but which at the same time is "fair" to the companies, to the labor unions, and to the public's interest. As he is "outside" the system, so to speak, his purely rational concept of fairness may not coincide with the concept of fairness on the part of any of the parties who actually control the operations of the port. Consequently he may have to compromise his notions of fairness with those that he thinks will be accepted by the parties at hand.

The research team visualized the picture as follows: if the industrial engineers did succeed in designing technologies that led to some cost savings in loading and unloading vessels, then these cost savings should be spread between the companies and the labor unions in such a way that the public's interest was maintained at the present level. Consequently, the problem narrowed down to determining some formula of advantage for the companies and the labor unions. The problem, however, could not be pinpointed better until the research team had succeeded in describing the port as a system.

From the point of view of the scientist's candidate for the decision maker, much of the situation that occurs in the port must be regarded as the "environment" of the system. First of all, there is the shipper's policy. This policy generates the schedules by which the ships arrive and depart from the port. The decision maker has no control over this policy, and consequently it has to be taken as a given. Note again the importance of the decision on the environment of the system; if the schedules could be changed, then it might be possible to arrive at a far smoother operation of the port. As it is, the study team accepts the schedules, however they are designed by the companies. A more general comment is that the entire mode of transporting goods may be wrong; e.g., perhaps trucks and airplanes should make ships obsolete. Once the study teams elect to take the schedules of ship arrivals and departures as givens, then they also assume that the existing mode of transportation by sea is a sufficiently large system for them to examine.

Given that the existing schedules are the appropriate basis for estimating demand on the port facilities, the problem is to determine what the demand actually is. One way to solve this problem would be to examine every shipper's schedule and try to determine exactly when his ships arrive and depart from the port and what he will be loading and unloading. This solution, however, would be very costly and would also result in many kinds of inaccuracies because shipping schedules obviously are not followed rigidly.

It seemed preferable, therefore, to examine the shipper's policy of scheduling ships in and out of the port by examining a real port and determining the frequency of arrivals and departures. The Port of San Francisco was selected for this study primarily because, in addition to its pleasant climate, the port is of medium size and there is virtually no seasonal variation. The scientist's idea was that this fairly simple port study could form a "prototype" for other port studies.

The researchers examined past records of arrivals of vessels into the San Francisco port and determined, for example, that the average number of vessels arriving during the day on a weekday was eight, and that the number plus and minus around this average was distributed according to what is called a "Poisson distribution." This fact immediately suggested to the researcher's mind that the port could be regarded in terms of a waiting-line model, similar to the one discussed earlier in connection with the arrivals of aircraft and the arrivals of calls at a central station of a telephone company. The arrivals of customers for service follows not a determined schedule but a probability schedule, and very often the probability schedule is Poisson. In the Poisson distribution, most of the arrivals are near or at the average, but there will be days when fewer than average occur, and days when there are far more than the average.

Once one begins to look at the system as a waiting-line system, the natural question is to ask how service occurs. Now the service of vessels in a port consists of a gang of men working in a hold and either loading or unloading the vessel by means of pallets which are hauled over the side of the vessel by a winch. Each gang has about 14 men. The unit of work therefore is one gang working for one shift. To describe the service of the ships one determines how many gang-shifts are engaged in servicing a vessel during its stay in port. Some vessels require only one or two gang-shifts and others require as many as 120. Obviously the vessels requiring larger numbers are more important in terms of the description of how the port operates, but the smaller demands also must be considered in terms of the allocation of the labor force.

Here again the distribution of service effort for each vessel followed very closely the "classical" lines of the telephone studies. In the telephone studies the telephone company determined how long it took to service each customer when he called in. In some cases it took an exceedingly short time because the customer knew the number and the operator could connect him directly, whereas in other cases it took a long time because the customer was uncertain about the number, or the connection could not readily be made.

In the case of the loading and unloading of vessels, however, there is an additional complication. The ship owners do not load and unload at a constant rate. For various reasons, at one shift they may engage six or seven gangs, whereas in another shift, e.g., the night shift, they will engage only three or four or even none. Consequently, one has to determine the statistics of the loading and unloading of the vessels in order to get some notion of how long a vessel will be in port and what kind of demands it will make on the labor supply.

Note that in this illustration the research team relied heavily on past experience as a basis for determining how the system works. This tactic on their part should be used only when the research team has concluded that the past policy in effect is part of the *environment* of the system, i.e., part of the system that is *not* controlled by the decision maker. For example, from the scientist's point of view, it would be quite incorrect for a company to use past sales data as a basis for estimating demand for their product in the future *unless* the company had for other reasons decided not to try to change demand by means of advertising, price, or technical improvement of the product. It is quite possible that the future should not be like the past if the demand can at least in part be controlled by the decision makers. In this case, however, the demand is not controlled by the scientist's candidate for the decision maker, and consequently the use of past statistical information is warranted.

The researchers had now arrived at the following description of the port. They could forecast in probability terms the arrivals of vessels on weekdays, Sundays and holidays, and the like, and they could forecast on a probability basis how many gangs a given vessel would request at each shift. Consequently, they could forecast for any given day the number of gangs that would be requested in a given shift.

One final aspect must be included: the availability of labor, a factor that changes from shift to shift because of illness, vacations, or just the personal wishes of the workers. Again, past statistics could be used to forecast the probability distribution of available gangs for typical shifts (weekdays A.M., P.M., Sundays, holidays, etc.).

In order to make all of these forecasts fit together, a computer simulation was developed. In the simulation a random number is picked which determines how many ships arrive, say, on a Monday morning. For each ship arriving a determination is made on a probability basis of the number of gangs the ship will require and the ways in which these requirements will be spread out over the days that the ship is in port. Other complexities, such as the ship's going into dry dock, can also be inserted into the simulation.

The simulation also picks out a random number that tells how many workers show up at the hiring hall and hence how many shifts are available. On some simulated days there are not enough gangs to service the ships, and some ships have to be idle until the next or subsequent shifts. On other days, there isn't enough work to go around, and the workers are idle. Thus the simulation forecasts the likely costs of labor shortage (idle ships) and work shortage (idle labor). In effect, the computer "acts out" how the port behaves. The great advantage of the simulation, of course, is that now one can begin to make changes in the simulation without in the least affecting the real system and so determine how a change in policy, e.g., an increase in technical efficiency of loading and unloading, would affect the port.

The next step was to see whether or not the simulation was realistic. This can be done on a crude basis by determining whether what the simulator does is similar to what the real system does in terms of demands for the labor force, the arrivals and departures of the ships, and the like. I say that this is a crude method of checking simulation, but it is often quite essential. Sometimes researchers unknowingly will put into the simulation various conditions that make the simulator "blow up." In this instance, the blowup might mean that in the simulation there was an increasing number of ships waiting to be serviced without the gangs available to serve them; the simulated situation would get increasingly worse, whereas in reality nothing of the kind occurs at the San Francisco port, unless there is a strike. Luckily, the simulation turned out to be reasonably close to reality.

Then occurred an event which helped the research team considerably in increasing their confidence regarding the simulator. Shortly after the study had started there was an economic recession which resulted in a decrease in the amount of cargo handled by the maritime industry by about ten to 15 percent. Unfortunate as the recession might have been from the shipowner's point of view, it was quite fortunate from the researchers' point of view. The decrease in cargo handling could be regarded exactly as though a technological improvement of ten to 15 percent had occurred. That is, if the industrial engineers were successful in increasing the efficiency of cargo handling, then this would have meant that, say, ten to 15 percent less work force was required to handle the cargo. From the model's point of view, the economic recession and the technological improvement, therefore, could be regarded as exactly the same kind of phenomenon. Now the research team could forecast by means of their simulator, without yet looking at the effects of the economic recession, how the port would operate if the demand for the handling of cargo went down ten to 15 percent. This could be done simply by changing some of the conditions of the simulation, e.g., by decreasing the average number of gangs required by ships when they were in port. Here again the research team was lucky; the simulation turned out to match the realities of the economic recession quite closely.

Of course the astute reader will recognize that, if the work load drops ten percent, and all the work is eventually done, the effect on the labor force is obvious: it will have ten percent less work to do and no elaborate simulator is required to tell us so. But it was of some interest to determine the likely patterns of idle times and waiting times; furthermore, if the labor union shrinks by attrition, what is the average waiting time apt to be? Finally, if one can score the system effectiveness, how can the optimal system be determined? These are questions whose answers required the subtlety of a computer simulation.

In addition to the simulation effort, the research team thought it advisable to see whether or not it was possible to develop a mathematical model of the operation of the port, using some simplifying assumptions. This is always an excellent idea when simulation is used because (1) simulations are expensive and (2) it is sometimes difficult to interpret all the results that are printed out by the computer. Wherever possible, a simplified mathematical model as a backup of a simulation is advisable.

The logic of the mathematical model is easy to grasp and illustrates very well how management scientists think. Begin by thinking of all the ships that are in port at the start of a shift. Each ship will have a remainder-of-work to be performed before it leaves, either loading or unloading. The work-to-be-done is measured in terms of gang-shifts. Call the total work-to-be-done W. At San Francisco on a normal day W might be something like 350 gang-shifts. Of course only a part of W will be accomplished in any one shift.

Prior to the start of a shift, some ships have arrived, each carrying its own total work-to-be-done. These arrivals add a certain amount to W—call it A—so that the total work-to-be-done is now $W + A$. But A is not a constant, for on different days the amount of work-to-be-done by arrivals will vary according to a probability distribution. Thus $W + A$ represents such a distribution and can be determined from past data by means of statistical analysis.

Now the shift begins and the labor force comes to work. The question is, how much of the work-to-be-done—$W + A$—do they accomplish? The answer depends in part on how the shipowners schedule the work and in part on the way in which the workers show up at the union hall. Assume, now, that the larger $W + A$ is, the more work will be done in the coming shift. This assumption suggests to the statistician that he use past records to estimate the work done in a shift—call it S—as a function of the total work-to-be-done—$W + A$. At the end of the shift, the total amount of work remaining to be done is $W + A - S$; this is also a statistical distribution, since both A and S vary from day to day.

Now, in the spirit of Sisyphus, we begin all over again on the following shift. There is $W + A - S$ work to be done, and a new arrival of work, and another try at reducing the total. By reiterating the calculations, we can eliminate any error in the original W, and converge on a steady-state, the endless distribution of work-to-be-done. This steady-state description enables us to predict the probability of waiting ships and idle work force. Furthermore, changes in the system of the sort described above can be calculated for their effects on waiting time and idle work force. Note that in the mathematical model the details of the work allocation at each shift have all been eliminated by statistical averaging; this means that the model is simpler and hence less likely to predict as accurately. But the simulation and the model together increase the scientist's confidence in his method, provided they essentially agree, as they did in this illustration.

The research team assumed that the components of the system were the companies, the registered members of the labor union, the casuals, and the public. The simulation determines what a technological change will mean from the point of view of each one of these components by predicting how a technological change will affect their objectives.

Because the companies and the labor unions are organized units, but the casuals are not, the research team decided that its major effort would be to study how a technological improvement should be divided between the companies and organized labor. The reader will see right away how this kind of a description of the system leads to the "embedding" principle that I have been emphasizing all along. If policies are to be set up for dividing the benefits of technological improvement between organized units of the system, then obviously the broader policies of other ports and even the nontransport industry will come up as issues. Is this an equitable way of spreading the gains of technological advance? Note that the study of technological improvement was supported by the United States government, which means that it was being supported by the public. There is, therefore, the reasonable question whether the advantages accruing from government-supported technological research should not be spread more widely across the public rather than directly to the companies and the labor union. There is also the question whether the casuals, who are the people who benefit from the port operation but are not union members, should not also have their interests represented in technological improvement.

Finally, as in all of these stories, it is interesting to see what the immediate result was. In the process of the study the companies and the labor union came to an agreement on the manner in which technological advances should be divided. One of the companies itself had been introducing a new technique of handling cargo and this by itself brought about a basis of agreement with the union. The basis of agreement did not drastically differ from that recommended by the research team, but certainly suffered from all of the difficulties of justification that I have just mentioned.

In this illustration one finds the same general theme that will run throughout this book, namely, that on the one hand there is more of a systems approach when one looks at the entire port rather than at just efficiency improvement in the handling of cargo, and hence there is a good basis for saying that the scientist's systems approach was a sound one. On the other hand, there is a higher level in terms of which the outlook of the research team was too confined, namely, the whole question of the distribution of the effects of technological improvement on society. Who shall benefit and how shall the inequities that result from technological change be handled?

From the management scientist's point of view, however, the criticism of his work represents no serious difficulty; in order to proceed in a feasible manner, he says, each step must be taken in turn. Once a series of studies of the kind just illustrated is made, then one is in a far better position to determine the principles that should govern technological utilization across the nation or throughout the world. Naturally we will want to keep reexamining this claim of the management scientist. As I have said before, it may be a piece of arrogance on his part to demand that we be patient and wait until all his steps have been accomplished.

The study did show that a subsystem of the nation's transportation system, namely, the loading and unloading of ships, is a feasible area of study by means of the scientist's systems approach, so long as one is not too fussy about the system's measure of performance. Indeed this and similar studies have become prototypes for descriptions of terminals in other areas of transportation, airports, railroad and trucking terminals, etc.

The study also shows how models and simulation play their role in the scientist's thinking. In order to examine this role more thoroughly, we turn to a very useful and important type of model.

5. INPUT-OUTPUT

We have just examined an illustration of the management scientist at work. Now we turn to an examination of the basic logic which he uses, which he calls a "model" of the system. A model, for the scientist, is a way in which the human thought processes can be amplified. As we shall see, this method of amplifying and making more powerful the thinking process takes the form of models that can be programmed into computers. At no point, however, does the scientist intend to lose control of the situation simply because he gets the computer to do some of his thinking for him. He controls the basic assumptions: the computer derives some of the richer and more complicated implications.

I shall be illustrating the model-building process in one way. Management scientists have found it quite useful on occasion to think of a system as a recognizable kind of entity into which are "inputted" various types of resources (people, money and the like), and out of which comes some kind of product or service. When we do think about systems in this manner, we arrive at what is called the "input-output approach" to systems.

Consider, for example, the educational system of a state. The legislative body "inputs" money and out come students with various kinds of degrees, high-school, college, and graduate. In this process the input is transformed into buildings and teachers, administrators, books, etc. And the transformed inputs then process people who are called students, who come out of the system with various types of education and training. It is interesting to note that, when we look at education in this manner, some of the components of the system—for example, the teachers—are both an output of the system and also then become one of the inner processors of the system. That is, the system creates some of its own potential.

Another example is transportation. Here the input is money and various types of materials and the output is the transportation of people and goods from one destination to another.

The input-output approach is an excellent way of looking at the industrial firm. The firm's input can be regarded as the initial investment of funds, and out of the investment come various kinds of products distributed to various consumers, as well as dividends returned to the investors.

Now we could think of the system as "a black box," as many people do. In this case, we would merely ask what output the system produces for a given mixture and amount of input. We would not be concerned with the "innards" of the system, that is, the way in which the inside of the system operates on the input to transform it into the output. In this case the management of the system is mainly concerned with the mix and the amount of the inputs; they are trying to manage this input so as to maximize some desirable amount and quality of the output.

But in this chapter we want to explore the inside of the box. We want to find out what kinds of activity should go on inside the system in order to produce the most satisfactory kind of output.

To this end we turn back to the list of five considerations given in Chapter III that are the management scientist's basis for a system.

To recapitulate, we need to develop a *measure of performance* that is to be maximized. We note right away that the total *amount* of output in all likelihood is *not* the performance measure. We might be quite impressed by the number of students who come out of educational institutions, or the number of cars that succeed in getting through the freeways, or the number of products that a firm makes. But our first impression has to be modified somewhat by the cost of this effort as well as the quality of the output. The cost in general will be measured in terms of the input. The quality has to be measured in terms of some ultimate objective. Therefore the measure of performance for a system will be something like a weighted output minus the cost of input, where the weights are determined by standards of quality.

The *environment* of the input-output system is the set of conditions that are relevant to but not under the control of the managers. In part, these are expressed as constraints on the system: the outside limits on the resources that are available, the characteristics of the demand for the output, and the like. In part, too, the "environment" describes the technology of the system—i.e., the manner in which input becomes transformed into output, provided that research and development are not "parts" of the system. If they are, then of course the level of technology may become a consideration of the managers.

The *resources* of the input-output system are the basis for the input, usually dollars, personnel, and equipment. The managers must decide how much of each type of resource to make available; once they do so, they set constraints on the activities of the system, e.g., in terms of the dollars each component can spend, or the personnel each component can employ. But, unlike the environmental constraints which are taken to be given or fixed, the resource constraints can be changed by the managers, e.g., by borrowing more capital, or by reallocating resources among the components.

Next there are the *components* of the system. We shall want to talk about the components in terms of various kinds of "activities" that are performed within the system. In the case of education, this is the *number* of teachers or man-hours spent in teaching, the *number* of classrooms available, the *number* of books in the libraries, and so on. As I said, the management scientist is not apt to be impressed by the sheer size of an activity, unless it can be shown that the activity is a critically important one. The scientific approach to the system consists of relating the amount of each component activity to the measure of performance, i.e., the output score.

Finally we will want to look very carefully at the *management* of input-output systems in terms of the control of such systems.

In order to make the management scientist's ideas clear, we are going to have to ask for the reader's indulgence a bit; he is to allow a good many considerations to become idealized and simplified in order to highlight the basic logic of the input-output analysis. In the next chapter we'll examine a more realistic situation that goes to the other extreme because its complexities and uncertainties are so great; many of the ideas developed here will have to be modified in order to fit the realities of this example.

For the present, consider a manufacturing firm that makes 100 products, say 100 different kinds of furniture. Suppose that this firm must sell each item it makes at a fixed price (this is already an unrealistic assumption, but I've already asked for your indulgence, so be patient).

Next, let us assume that the accountants can determine for us exactly how much it costs to manufacture and distribute each product (this is also unrealistic, as we shall see). These two suppositions on our part will enable us to say that for any given product we can determine what kind of profit it generates for the firm. As we think about the firm from the point of view of its investors, we'll say for the moment that the main objective is to maximize net profit, i.e., gross return minus costs. Obviously, all we need to do is determine how much money a given product returns to the firm and its investors, and subtract from this the cost of its manufacture and distribution, and we come up for a given year with a total net profit associated with a given product.

In this very stripped-down example, then, we can readily identify the five steps of system description:

1. The measure of performance is the net profit.

2. The *environment* is the constraint on capital, the price of each product (note that this was *fixed* by our assumption), the demand for the product (we assume no action on the part of the managers can change this), and the technological level (the number of items of each product that can be produced per unit of resources).

3. The *resources* are the dollars and personnel of the system.

4. The *components* are the product-lines, i.e., those subsystems that produce and market each product (note that in this simplified example, the identification of a component—or "mission"—is more or less obvious).

5. The *management* is the decision making on the amount of resources to make available to each product-line.

We turn our attention to the components, as these are basic to our design of the system. We say that the number of items of a given product that we produce in a given year is the measure of the activity associated with a product-line. Each product activity represents a component of the total system, and associated with each amount of activity is a number that represents the benefit to the firm.

Specifically, in this very simplified illustration, we assume that the contribution of each product-line to the total performance of the firm—i.e., its total net profit for the year—is solely a function of its own activity. The more activity, all other things being equal, the more the total profit to the firm. In other words, each product-line operates "on its own" and does not interfere with or improve the productivity of other product-lines. In systems language, it is a "separable" component.

However, we should note that some products may be more profitable than other products. This means that their contribution to the total net profit per unit of activity is greater than is the case for other products.

The situation can be represented quite simply in mathematical terms if we assume that the contribution of any activity to total profit can be stated in a very simple "proportionality equation." If z represents the total net profit of a firm, then, all other things being equal, we will be assuming that z has a specific relationship to each product-line activity. Hence if we let x_1 represent the number of items of product-line 1 that are manufactured and distributed throughout the year, then we assume that z has the following relationship to x_1:

$$z = a_1x_1 + b_1$$

For example, if each item of the first product-line sells at 200 dollars, and the cost of its manufacture and distribution is 150 dollars, then the profit z will go up 50 dollars per unit of the first product; so the a_1 in the equation above is 50. The b_1 represents what accountants call the "fixed costs" associated with plant, administration, and so on. These are costs that cannot be avoided so long as the company is in business. The fixed costs are distributed over the product-lines according to some accounting practice, so that b_1 represents the amount of fixed cost (burden) assigned to product-line 1. As we shall see, it does not matter how this distribution is accomplished, in relation to the design of the system. Thus, though accountants often fuss endlessly with what they call the "assignment of burden" (i.e., the b's) to each department, in terms of the scientist's approach to systems, the assignment is irrelevant.

If we proceed in this manner, then we have a very clear notion of how the components contribute to the entire company's measure of performance. If we turn to product-line 2, we assume that its contribution to the firm's performance is given by

$$z = a_2x_2 + b_2$$

Note that a_2 and a_1 need not be the same; if a_2 exceeds a_1, then the second product-line contributes more per unit of activity than does the first. This does not necessarily mean that the second product-line is better than the first, because it is possible, as we shall see, that the second uses up very scarce though possibly low-priced resources.

In general, we take all the equations, like the ones above, for each product-line and put them all together in one grand equation, which looks like this:

$$z = a_1x_1 + a_2x_2 + a_3x_3 + \text{----} + B$$

In this case, the a_1, a_2, and a_3 represent the net profit accruing from each one of the products, 1, 2, and 3, and the B at the end of the equation represents the total fixed costs of the firm, i.e., $b_1 + b_2 +$ ----. It now looks as though we have arrived at a very simple way of managing the system. Our problem is to determine the amounts of each activity, i.e., x_1, x_2, x_3, in such a way as to maximize z, the total net profit of the firm. We can immediately see that the maximum of z is completely independent of B, the fixed costs, as well as independent of the way B is allocated to the product-lines.

But now something odd seems to have crept in to our considerations. We note that x_1, x_2, and x_3, which are activities, cannot be negative. That is, in the scientist's thinking it is impossible to produce negative amounts of items. Therefore, all of the x's are either zero or positive. Hence z is bound to increase as x increases, and this would suggest that the more activity we carry on the better, so long as the a's are not negative. But if an a is negative, this means that a given product sells at a price that is less than what it costs to make it. In such case the managers have an easy decision, namely, to stop making that product. So we can assume that the *a*'s are positive, and hence that they are greater than zero, for if they were zero, then clearly again the firm should not make the product (it would take a loss because of the fixed costs). If the *a*'s are positive and the *x*'s are positive, then *z* is bound to increase as we increase the amount of activity. We note, in fact, that, in general, if the firm cannot do better than assign zeroes to all of the *x*'s, then it should go out of business, because then it's just simply maintaining buildings and administrative force at a direct loss.

Therefore, if our objective is to maximize *z*—the measure of merit of the total system—it seems as though we ought to carry on as much activity as we can, and especially activities associated with the most profitable products.

But it is at this point that the two other critical items on the list—namely, the resources and the environment—become important in our considerations. For example, some of the pieces of furniture may need skilled labor, e.g., to carve out various kinds of ornament. But the firm may have only a limited number of people who can perform such skilled tasks. Hence there is a limit on the number of items of certain products that can be manufactured given the existing labor force. This example should help to clarify what I said earlier about a profitable item's not necessarily being better than a less profiable one; the more profitable one may use up skilled labor, even though the actual price paid for this labor may not be excessive. Thus even though a product-line may be profitable, it may still require so much skilled labor that only a few items can be made.

Again, it may happen that some of the products need a special kind of wood, and the firm has only a small supply available. Or some of the products may need special equipment. And so on. These are what in Chapter III we called the "environmental constraints" on the firm. One basic constraint is the upper limit on the total amount of capital that the firm can pour into the manufacture of its products. It cannot make all of the *x*'s very large, because it doesn't have the money to hire the labor and build the equipment to do so.

But these considerations which we have been talking about in the English language are easy enough to translate into mathematical language. If, say, some tables and bureaus and chairs need skilled labor, then all we need to do is say that the number of products of each of these kinds that can be manufactured is limited by the "environment," i.e., the total amount of skilled labor. Suppose, for example, that t_1 hours of skilled labor must be spent in making one unit of product x_1, and t_2 hours of skilled labor must be used in making product x_2, and t_3 hours of skilled labor must be used in making product x_3. And suppose that these three products are the only ones that need this particular kind of skilled labor. For example, product 1 may need three hours per unit, and product 2 may need five hours, and product 3 may need eight hours. Suppose that the total number of hours available is 1,500 per year. The whole situation is quite simple to express in mathematical terms:

$$3x_1 + 5x_2 + 8x_3 \leqq 1{,}500.$$

The manager's problem has now been translated by these considerations into a mathematical problem. For each limitation that the environment imposes on the system, there will correspond an equation of the type we have just expressed. For example, the total amount of money available becomes translated into the total amount of dollars that each product eats up as we manufacture varying numbers of units of it. And this total amount could be expressed in the following sort of equation:

$c_1x_1 + c_2x_2 + c_3x_3 + \text{----} + c_nx_n \leqq C$ (the total budget allowed for the system minus fixed costs).

From the management scientist's point of view, the manager's problem has become translated into a mathematical problem of how to maximize the profit equation given above, subject to the environmental constraints, which themselves are expressed in terms of mathematical inequalities.[1]

If the number of products is small and if there are some rather obvious relationships between their profitability and the total profit of the firm, then it may be quite obvious how the manufacturing firm should proceed. In a systematic way we would examine each product and its contribution to the total profit and, by a bit of manipulation here and there, would try to fit the manufacture of the most profitable products within the environmental constraints. In the end we would arrive at a managerial plan that comes very close to maximizing its total profit.

In fact, in fairly straightforward cases it's amazing how close managers have come to what the mathematicians deduce when they use the more precise methods of mathematical analysis.

However, it's easy to see that the problem can get quite complicated and that no commonsense approach will be sure to arrive at the correct answer. As the number of constraints and activities increases, the human mind is simply not capable of sorting out all of the various kinds of situations that might apply and putting the pieces of the jigsaw into the right places.

In the last two decades a great deal of work has been done on problems of the type we have just been discussing. There are straightforward ways of maximizing the net profit, expressed by the equation given above, subject to all kinds of constraints that are imposed by financial interests, by technologies of various kinds, by personnel limitations, and so on.

In fact, in some cases, if the activities can be broken down in certain ways, it may be possible to handle over a million activities subject to something like 35,000 constraint equations. It's obvious that a lot can be described in these terms. Naturally in such a situation it is essential to use computers to carry through the mathematical operations by what mathematicians call an

[1] The mathematics is usually more complicated than this example indicates, since the utilization of different kinds of resource by each component needs to be considered.

"algorithm." The manager in this case puts into the model the basic information about net profit, costs, and constraints. The computer, aided by the mathematician, generates the "optimal" plan.

Models of the type we have been discussing are often called "linear-programming models." The label "programming" seems clear enough, because the type of thinking we have been examining has to do with the way in which activities should be programmed. That is, we have been thinking about the amount of activity (including none at all) that should be carried on by each component of the system. The term "linear" describes the way we have set down the equations and constraints. In each case we have made the assumption that the profit return, costs, time, and the like go up proportionately to the number of items that are produced and distributed. In mathematical language, all of the equations that we have set down are "linear" equations.

There are some very beautiful things that can be done with linear-programming models. These things have much to do with the way managers typically think about their systems and especially about their resources.

The basic distinction between a resource and the environment is that managers can control the size of the resource, whereas the environment is given. The distinction is not an easy one to make. In fact, one criticism that is often leveled at managers is that they are far too rigid in their thinking about constraints; they may act in too cautious a fashion by assuming that they can't go beyond certain "prescribed" limits.

For example, we talked above about the fact that certain products demand skilled labor. If some of these products are quite profitable, the bold manager may wonder whether he should not hire another man to do the job and thereby double the amount of skilled-labor time available. This, he would argue, would permit him to make more of the profitable items. A more cautio s manager may feel that he has gone to the limit on hiring skilled labor and so refuse to budge beyond the constraint. In the first case, the manager regards the additional skilled labor as a resource, whereas in the second case the manager takes it to be environmentally limited.

The scientist can help in this debate between the cautious and the bold. By a mathematical technique he can help answer the question by showing the cost associated with a given constraint. That is, he can show what profit would accrue if additional hours of skilled labor were made available. This technique, he says, is quite useful in setting the budget for the firm and is highly informative for the investors. The budget, in effect, sets the limit of expenditures in various components of the system; one thing that plagues the mind of every budgeter is whether or not the limitations he has set are too liberal or too strict. A linear-programming model can tell the advantage that would accrue from adding or subtracting dollars assigned to each activity. But elegant as the technique is, it does not completely solve the investor's or budgeter's problem. If I learn that, by relaxing a constraint, I can increase profit by so many dollars, then what? As an investor, I have to decide whether this increase is warranted in terms of other opportunities for investment. But this means that there is a larger system—the whole spectrum of investments—which I must consider, but *which the management scientist has ignored.*

We see again the kink in our hero's armor: he seems tied to the particular system he is studying and must allow some other form of "nonscientific," or even "nonsystematic" judgment to take over when the issues go beyond the system to which he is shackled. The master in this act is the investor, who scans the whole world of investment opportunity; the management scientist is his slave who works in one of the master's many dominions.

The same point arises when we ask who shall have responsibility for the data that the management scientist uses. It is certainly not easy to determine how much it costs to make a product, even in the most simple case. At the outset I said that to assume the accountants will give us the right data is unrealistic. One cannot always believe what they tell us because some of their work is done for taxation purposes, and hence the cost determinations may not really be relevant to the profit side of the picture. In any event, uncertainties about prices and sales and so on introduce the need to make various kinds of subjective judgments, and when one inserts these into the model one may feel that the inexactness of the information might produce totally wrong results.

But a deeper issue arises from the same basic theme that surrounds the determination of the proper constraints: each dollar that a product-line uses is a dollar of the investor. But the investor is interested not only in the 100 product-lines of this firm, but in a multitude of other opportunities as well. Hence, for him a "cost" is to be gauged in terms of all his lost opportunities. The accountants, however, are concerned with just this system and its 100 product-lines. Their narrow vision may totally distort the true cost to the investor. And if the scientist accepts the accountants as the data collectors, he too is unreliable.

Furthermore, we have pointed out how unrealistic the model of the firm may be in terms of its "simplifying" assumptions, which were introduced in order to attain mathematical elegance.

To the charges of using the wrong data or an oversimplified model, the scientist can make a partial reply. There are, he says, some elegant mathematical "sensitivity" techniques that provide a way of determining the seriousness of inaccurate sets of data. It's possible to ask the computer to run an experiment in order to answer the following question: What would the result be if these costs were twice as high in reality, or what would the result be if the assumptions were more realistic? But even so, if the possible error is serious in terms of the investor's interests, then what?

The obvious reply by the scientist is also a rather difficult one for him to make, namely, extend the model into the more complicated and larger considerations of the manager. Indeed, to some extent the scientist can do this. For example, he can handle the problem when the relationship between the amount of the activity and the contribution to the measure of performance is not linear. This might occur, for example, when the quantity sold depends on the price, so that the lower the price, the more items are sold. In the simple model, we did not go into detail about that critical aspect of the environment which we call the demand on the system. We didn't do this because the demand on the system was, so to speak, infinite. According to our very simple assumption, the consumers would buy any amount of furniture at the price offered. The more realistic situation is one in which consumers react negatively to high prices and react favorably to lower prices. If advertising is put into the picture, then other complications occur. We need to spell out within our model the various ways in which consumers can react to the things that the system does and most of these reactions will not result in simple linear types of equations. However, in many instances it is possible for the management scientist to express the relationships in nonlinear terms and to solve the problem, even though the mathematics may become more complicated and the total amount of computer time greater.

Furthermore, it's often worthwhile to look at a system from the point of view not of certainty, but rather of probability. If probabilities rather than certainties are assigned to the events that might occur, then the scientist can use the very powerful techniques of probability theory in order to solve the manager's problems.

There is also the size of the mathematical model. It may be true that not all the investor's opportunities were captured in the simple model of the 100-product-line firm, but in time we can enlarge this model to include more and more of the investor's world.

The enlargement of scope and complexity in management science is not too different from that in any field. Modern industry began with very simple kinds of models of machinery; as time went on, the engineers determined how to make the machine do all sorts of complicated things in complicated environments. In the same way, the mathematicians who are working on management models are able to put into the models all kinds of considerations as time goes on. They can enlarge the size of the models, put in various uncertainties and judgments, and in fact begin to approach the realities that face the manager.

There have been many applications of the linear-programming approach to the study of systems, production, distribution, marketing, personnel assignment, and so on. In these applications the scientist is not so rigid as the model might indicate. He realizes that there is plenty of room for error, and hence the need to design the management part of the system so it can react in a timely manner to the consequences of error. Especially, it is essential to avoid misunderstandings in the organization about what the model can and cannot do. There is one famous case in which linear programming was "successfully" applied to the allocation of freight cars of a large railroad. A computer installation was formed which printed a daily schedule, derived from the data that was fed into it about customer demands. But, although top management understood this systems approach, middle management did not. Middle management believed that either the computer was right, or they were. As the computer had had little or no experience with railroading and its complexities, the choice was obvious. Instead of regarding the model as an aid to daily decision making, they took it to be an ignorant and dangerous monster. And they had the advantage, because they fed the monster its daily pabulum of data; what more simple than to feed it stale bread—data that are a week old, say? There's nothing sillier than a computer that spills out week-old schedules. The costs of computing were the same, silly or not. The result was that the scientist's model "didn't work."

To the scientist, this example may say nothing except that one should always pay attention to the attitudes of people. But there is a deeper implication here. For one thing, the example shows the mistake of assuming that, simply because the logic is correct, the "solution" is also correct. More generally, it shows that there is a fundamental limitation of any modeling of a system, that *the system is always embedded in a larger system.* From the point of view of the investor, any given company is only a part of the total range of his concern. From the point of view of the middle managers, the company is also only a part of their whole life's concern, and when ideas are generated that threaten their "systems," they naturally resist them.

Consequently, no matter how marvelously a specific system works, in terms of a larger system it may not "work" at all.

In sum, the true costs associated with any system always reflect the way in which the larger system behaves. What the scientist has been calling the "systems approach"—at least at this point in our considerations—has always been the limited system. In order to arrive at measures of effectiveness, consideration of resources and environment, and the components, it's been essential for him to limit himself to assuming some measure of performance as acceptable that is meaningful in a limited sense only. I said above that the linear-programming models might be very useful to budgeters in considering how to allocate funds to various departments and divisions of an organization. As long as the budgeting operation is taken to occur within the system, there is no question about this advantage. But the larger system that allocates the funds in the first place to the organization cannot be described by management science unless the scientist is able to generate an even larger model to describe the activities of the larger system. However, as one moves to considerations of larger and larger systems, the problems of complexity become enormous. This can be seen most clearly in the case of the measure of performance of a system. If the scientist accepts from some authority that the measure of performance shall be net profit, or shall be the number of students graduated, or shall be the number of cars using freeways, then the problem of values doesn't appear in any critical form. It doesn't appear because some other, larger system has decided what it shall be. But if now he tries to look at the larger system, he has to ask himself whether these measures make any sense, and here his guidelines begin to become quite fuzzy indeed.

In general, we can say that the larger the system becomes, the more the parts interact, the more difficult it is to understand environmental constraints, the more obscure becomes the problem of what resources should be made available, and, deepest of all, the more difficult becomes the problem of the legitimate values of the system.

Nor can the manager accept with any easy confidence the notion that the scientist will gradually enlarge his perspective to greater and greater models, because the manager has every right to ask what system guarantees the scientist's success. After all, it was the scientist who introduced the "systems approach" in the first place. As he did introduce it, his opponents have a right to throw his own method back in his face. Is *he* a part of the "whole system"? If so, what kind of a part? Is there some characteristic of the whole system that says that science must succeed, even in the face of all the errors—irreversible errors sometimes—that it commits? As the scientist is urging us to examine "programs" for their "pay-offs," what about his own program? It's expensive, difficult to understand, and, according to his own confession, limited in scope.

But this attack on the scientist may be entirely too harsh. There is, in fact, a perfectly logical way in which he can defend himself by modifying his somewhat rigid position. After all, he may say, I am defending a philosophy of life as much as a method of learning. If my precise method of learning cannot yet be extended in a precise manner into the obscurities of larger systems, at least my philosophy can. And if my philosophy precedes my precise method in the exploration of whole systems, then, when my technical capability catches up, we shall be so much better prepared to use it.

What is this philosophy? Why, it's the belief that management systems can be looked upon as essentially information-processing systems, in which the information takes the form of data about objectives, environment, resources, and components (missions). Even though a precise model cannot be constructed, the "mode of thinking" that is inherent in programming models can be utilized in a very rich way.

The scientist's opponent will have a ready reply, or set of replies. The anti-planner can never feel that any "mode of thinking" that ignores rich experience and judgment can be any good. After all, he'll say, this country has always relied on the experienced man to lead the way, and look where we are today. The humanist, on the other hand, does look at where we are today and is not very happy about it. Apparently our "experienced leaders" are quite capable of making a mess of things in cities, countries, and the world at large. But the humanist is not very happy about handing responsibility over to characters like the scientists who think of men as information processors.

But we agreed to give the "scientist" enough rope to hang himself. I put quotes around his name because he's just made himself into a philosopher, and no one knows whether such creatures are scientific. In the next chapter we'll see how the scientist, turned philosopher, approaches the complexities of programming in the real world.

II. Applications of Systems Thinking

6. PROGRAM BUDGETING

We leave now the clear and untroubled waters of the mathematical model and try to traverse the muddy swamps of reality. But the management scientist wants to keep his logic, even though he may lose his equations. The "reality" we shall face is the problem of budgeting activities in a state government agency. This reality apparently defies precise formulation in terms of a model, but nonetheless the scientist will want to think about the problem in his way. The thinking we shall explore is frequently called "program planning and budgeting" (PPB). There are many definitions of program planning and budgeting in the literature of the subject, and most of us have become quite frustrated in our attempts to get a clear-cut and unambiguous description of the method. Obviously there is no such clear-cut, unambiguous description, because each author brings his own background into his definition. In this chapter we shall be considering the management scientist's notion of PPB.

Chapter II discussed a philosophy of management in which the manager's sole concern is to make sure that specific parts of his organization operate as efficiently as possible. According to this philosophy, once the manager has made up his mind that a part is necessary, then he should install the administrative details that will make that part function as efficiently as possible. Consequently, he will try to set the budget at the level which he believes is required for efficient operation. This way of viewing the budgetary process, says the management scientist, ignores the relationships between the different parts of the organization. It also ignores the relative importance of the operations of one part; as we shall see, it likewise ignores some other very significant aspects of organizational activity.

In order to get an overview of the system, we need to proceed, as in Chapter III, to look at the overall objectives, and then identify environment, resources, and components (submissions).

The reality we intend to examine is the problem of alcoholism as it occurs within a modern society, and specifically the problem of what a state government should do about it.[1]

Every one of us has been confronted with the problem of alcoholism in one form or another, e.g., drunk drivers, alcoholic relatives, a lowering of worker performance as a result of excessive drinking, and so on. Most of us are also aware of the benefits of alcohol taken in moderation.

As in the case of all the important problems of our systems, it is not easy to go from the vague feelings of discomfort we have about a problem like alcoholism to a very specific statement as to its true meaning, and hence what a state government should seek to accomplish in order to solve the problem.

An older generation is familiar with one solution to alcoholism that was adopted by the United States in the 1920s. The idea was to solve the problem of alcoholism simply by eliminat-

[1] The illustration examined here is taken from work done by A. H. Schainblatt at General Electric, TEMPO. My thanks to him for allowing me to use this material.

ing the sale of all intoxicating beverages throughout the country. The social benefits of alcohol were thus thrown out with its evils. That the designers of prohibition failed to have a systems approach became all too apparent to those of us who lived through those prohibition years. It simply was not the case that the legal prevention of the sale of alcoholic beverages stopped the consumption of alcohol. Indeed, in the upper classes there was real evidence that prohibition created a greater opportunity for alcoholism. The illegalities of alcoholic purchase produced an environment in which alcoholism could thrive far more than it had prior to prohibition. Prohibition also introduced a severe wave of organized crime. It is even doubtful whether prohibition succeeded in what was evidently its prime purpose, namely, the prevention of drinking among the working classes.

The spirit of prohibition as a solution to the problem of alcoholism can be expressed in the simple definition of the objective of an alcoholism "missoin": to reduce the amount of alcohol in the bloodstream of the average citizen. The idea of prohibition was that, by instituting legal constraints on the sale of alcoholic beverages, this quantitative measure of performance could be "optimized." The average citizen, it was believed, would consume less because he could buy less.

But is this objective—reducing the average amount of alcohol in the citizen's bloodstream—an appropriate formulation of the problem? To the management scientist, the proposed measure is similar to the proposal to "cut costs" in the efficiency philosophy; the aim, he says, should not be to minimize costs, but to optimize performance. Many people recognize the advantage of a certain amount of consumption of alcohol on social occasions. Even those who believe, on moral or health grounds, that no amount should be consumed are in no position to impose their ideas on the rest of the citizens. The trouble with the minimization objective is that it fails to express the real problems that are associated with alcoholism. The real problems lie in the kinds of behavior that are exhibited on occasion by people who have consumed what for them is too much alcohol. In vague terms, this means that the consumption of alcohol on the part of some individuals may create a social and psychological situation that is disadvantageous both to the individual alcoholic and to the other people with whom he associates.

We note that, in this definition of the problem, the term "creates" means that we must assign to the consumption of alcohol the specific social disadvantages that it produces. This might be done, for example, if we could gather data in the following manner. Suppose that we could assign to a given state a "gross product" of the state, i.e., a figure that represents the economic output of the state in a given period of time. Suppose we were able to relate this economic output to work-days on the part of the citizens of the state. Finally, suppose we were able to assign to the consumption of alcohol the number of days that were lost by the work force of the state (in this regard the work force includes both workers and salaried employees). If we were able to gather together all of these pieces of information, then we might want to say that *under the existing situation* the measure of the damage caused by alcohol is the loss in the gross product of the state that occurs from causes associated with the consumption of alcohol. I have italicized the phrase "under the existing circumstances" to emphasize the point that some redesign of the total system—e.g., the substitution for alcohol of some far more innocuous depressant—might completely change the whole nature of the problem. Hence, as always in management science, the particular measure of performance we select for a subsystem will change if the total system changes; at the outset of a project we usually restrict ourselves to the existing state of affairs in order to get started.

In this case, then, the suggestion is that the measure of merit of the alcoholism mission be the amount of alcohol-induced social damage that the mission has prevented, where "social damage" is measured in terms of lost days of work, minus the costs of administering the mission. We note that since lost days of work can be translated into *costs* of lost time, we may expect to find a common unit of measurement for the whole mission and its subactivities, i.e., a *net* benefit, which is a very desirable thing for the scientist.

The suggested measure of merit of the mission has much of the "nonhuman," if not "inhuman," attitude of the management scientist at work. The main point about alcoholism for the humanist is the heartaches of a family brought about by an alcoholic parent. Actually, many an alcoholic can earn a living satisfactorily enough, but his evenings of drunkenness produce the family tragedy. There are the moralists, besides, who see in alcoholism a breakdown of personal morality, which occurs whether or not the individual misses a day's work. Finally, the alcoholic housewife will not be counted in the proposed mission's measure of merit, because normally she will not be a breadwinner, and her unproductive days will be taken care of by other members of her family.

To all this complaining, the management scientist replies that there is no feasible way to get inside even a sample of homes with alcoholics, except for the poorer families visited by social workers; but these surely are a biased sample. Furthermore, how shall we measure heartache or immorality? As we shall see, the feasible becomes the archangel of the management scientist's heaven; he is fully guided in his practice and his morality by this spirit. In this case, lacking any feasible alternative, he lets his idea stand in order to guide his thinking, nonetheless keeping an open mind for other possible measures in the light of new evidence. For the start, therefore, he says that the problem of alcoholism under existing circumstances is the minimization of loss in the economic product of a state as a result of the consumption of alcohol. Even for the feasible-oriented management scientist, this definition by itself is incomplete, however, because it fails to recognize the benefits of alcohol for many of the citizens of the state. Hence the scientist seeks to determine a "safe" group of alcohol consumers, e.g., the purely "social drinkers." He then phrases the objective of the mission as follows: the minimization of social damage caused by alcohol subject to the condition that the opportunity of consumption by safe social drinkers remains the same.

With this preliminary definition of the problem accomplished, the scientist turns to the kinds of activities that are related to the measure he has formulated. Specifically, he wants to consider those activities which will affect—either positively or negatively—the social damage that is produced by alcoholism. In the language of program planning we will call these activities "programs." As a first estimate the programs can be represented in terms of (1) *prevention* of alcoholism, (2) *remedial* activities that take place when alcoholism has occurred, (3) *control* of alcoholism by means of medical, economic, legal, or social activities, (4) *research* on the problems of alcoholism, and (5) *administration* and general support of alcoholism missions.

These five general programs are quite broad; so in order to analyze the activities it is necessary to break the broad programs down into subprograms. For example, under *prevention* it is easy to see that (1) educational courses and lectures may assist the state in helping people to understand the significance of alcoholism, its dangers, and the means of reacting to signs of danger. The prevention program may also be carried out (2) by legal means, for example, by laws that are designed to prevent the sale of alcoholic beverages to minors, or the sale of alcoholic beverages during certain times of the day. And finally (3) the prevention program may be carried out by economic and financial methods, e.g., by taxes and price-control measures.

But one can also see that even in this breakdown there will be some overlapping—e.g., between the legal and the economic—and furthermore that it is not obvious that these three subprograms exhaust all the ways of preventing alcoholism. For instance, if alcoholism is thought to be a psychological problem, then psychiatric methods may be essential in prevention, as well as in rehabilitation programs when alcoholism is long advanced in the individual. But remember that the scientist has begun by restricting his attention to the existing conditions. Since in the total system of the state there is no program as yet for the prevention of alcoholism by means of psychiatric treatment, then he eliminates this possible subprogram in this phase of his thinking. He restricts his attention to the existing conditions so that he can arrive at a fairly precise evaluation of the current subprograms. As a result of this evaluation, he says, he will be able to look at the existing system and ask some very critical questions about it.

Turn now to the *remedial* activities of the mission. These activities are relevant to the cases in which alcoholism has occurred. Obviously the first subprogram here will be the detection of occurrence. As in the case of crime, the mere occurrence of the social defect means nothing unless the state becomes cognizant of it. Hence detection becomes a necessary subactivity associated with remedial action. Once detection has occurred and the state has become informed of an actually or potentially dangerous alcoholic situation, then the next step is screening and evaluation, i.e., attempting to assess the real dangers. This will be followed, in the case of severe alcoholics, by "short-term" treatment in detoxification clinics or hospitals, and then by a longer-term treatment, or rehabilitation. Rehabilitation might take place, for example, in clinics or in state mental hospitals. In some cases there will be no attempt to rehabilitate the alcoholic, but instead he will be kept out of society—in county jails or in care centers for prolonged stays. Hence we should separate this activity—call it "custody"—from rehabilitation.

Although the breakdowns of the broader programs into subprograms may seem more or less obvious, the management scientist has had a specific aim in mind, that is, to arrive at subprograms whose activities are quantifiable. Thus, he can determine the *amount* of detection effort in terms of man-hours or dollars, or the *amount* of treatment and rehabilitation in terms of patient-days. Indeed, in all these subprograms there are several choices for quantifying the amount of activity, and the scientist will seek to find that measure which most clearly contributes to the overall measure of performance of the system, the decrease in social damage.

In a similar manner, the scientist will break down the *control, research,* and *administration* activities into subprograms, each one quantifiable.

His next task is to determine how much of each subprogram is carried on in the existing system, not just inside the alcoholism agency, but anywhere in the state government, because it's quite clear that a number of the other parts of the state government play some significant role in the mission. Therefore, to determine the amount of activity carried on by the state in the alcoholism mission, the scientist must look elsewhere to assess the "alcoholism" activities in the other departments of the state. One can begin to see what the management scientist means by *program* budgeting: the programs and subprograms are identified across the whole system. So far in the discussion there is no budgeting, that is, no decision as to how to allocate funds and personnel to the various programs in order to carry on the alcoholism mission. At this stage in the thinking we are merely assessing how much of each program of the mission is carried on at the present time. This assessment can be made if we can develop a matrix of the type in Table 1. Along the top of this matrix are shown the various programs and subprograms of the alcoholism mission. The left-hand side of the matrix shows the government agencies that have some amount of activity in one or more of the programs. For example, the Department of Correction carries out a certain amount of remedial activity when it incarcerates an alcoholic prisoner; it detects, screens, evaluates, and on occasion may even treat and rehabilitate the alcoholic. As the existing system is set up, in other words, the responsibilities of the Department of Correction force it to play a role in the alcoholism mission. In the matrix, checks have been placed in squares to indicate those areas in which a given department of the existing system plays some role in the program and subprograms of the alcoholism mission, the absence of a check indicating that the department in question conducts little or no activity in the corresponding program.

Our task now is to replace the checks by some measure of

Table 1

AGENCY	PROGRAMS																	TOTAL
	PREVENTION	*Educational*	*Legal*	*Economic*	REMEDIAL ACTIVITIES	*Detection*	*Screening & Evaluation*	*Treatment*	*Rehabil-itation*	*Custody*	CONTROL	*Medical*	*Economic*	*Legal*	*Social*	RESEARCH	SUPPORT	
Department of Correction					√	√	√	√	√	√	√			√			√	
Dept. of Mental Hygiene					√	√	√	√	√							√	√	
Dept. of Social Welfare						√					√	√	√		√		√	
Alcoholic-Bev. Control	√	√	√	√							√		√	√				
Dept. of Youth Authority	√	√			√	√	√	√	√	√	√			√			√	
Dept. of Public Health	√	√			√	√	√	√	√		√					√	√	
Calif. Hway. Patrol	√	√									√			√		√	√	
TB Sanitaria Subvention					√	√		√	√		√	√						
Dept. of Motor Vehicles											√			√				
Dept. of Rehabilitation					√	√			√								√	
Board of Equalization											√		√	√				
Alcoholic-Bev. Cont. Appeals											√			√				
Dept. of Education	√	√															√	
Total	√	√	√	√	√	√	√	√	√	√	√	√	√	√	√	√	√	

activity. From the point of view of the budgeter, this means replacing the checks by dollar amounts, e.g., translating man-hours and equipment utilization into dollars.

Now the system problem becomes quite difficult to analyze correctly. Consider again the Department of Correction. What dollar amount is appropriate to represent the amount of "alcoholism" activity carried on here? Consider the following answer to the question: Count the number of prisoners who are alcoholic, compare this to the total number of prisoners, and apply the ratio against the total budget of the Department of Correction. Thus if ten percent of all prisoners are alcoholic, then the amount of alcoholism activity of the Department of Correction should be ten percent of the department's budget. But one could argue that this figure is totally unrealistic, because the Department of Correction is doing a great many other things with prisoners besides treating them for alcoholism; furthermore, not all alcoholic prisoners go to jail as a result of their alcoholism.

For the scientist, the question needs to be answered in terms of the significance of the amount of activity for the total system performance. An hour of a man's time that is spent solely on treating alcoholism should be assigned to some subprogram of the mission, wherever it occurs. If the hour of time also does other things, then only a portion of it should be allocated. Hence the scientist believes that *in principle* there is one correct figure (dollars or time) for each block of the matrix. However, when the scientist tries to determine this correct figure, he soon runs into political problems. If the figure assigned to the Department of Correction comes out to be large, then this department has reason to fear that its budget for the next year will be reduced because the legislators will argue that it is doing a lot of things that are not properly associated with its "correction" mission.

The situation may seem quite strange to the scientist who has not lived within the system. The assignment of numbers, which he feels should be an objective procedure, now turns out to be politically loaded. Of course, what is happening is that the larger system is intruding into the subsystem. The scientist wants to collect data to determine the proper budget, but the data themselves have meaning for all the relevant agencies of the government. Thus, politics reflects the extreme interdependency of agency activities, so that "alcoholism" as a mission is not really separable from other state missions. Because the scientist is in no position to take on the whole system at this point, he must instead make the assignment of numbers by a political negotiation, in which all sides are given the possibility of expressing the strongest possible argument for their position.

The scientist may feel a bit uncomfortable about this. He may believe that the systems approach is a straightforward and precisely rational method of reaching decisions about systems. He may think that political power struggles do not result in anything like an optimal plan for society, and that his systems approach is an attempt to remove all of the uncertainties of power politics and replace these by the "correct" method. But the scientist's need to be feasible prevents his creating a huge image of the system and plugging in what he thinks are objective data. The larger system is there, nonetheless, and its existence shows itself in political activity. The more astute scientist will recognize an advantage rather than a disadvantage in the politics of budgeting. In fact, he will come to see that his systems approach provides a novel and important arena for political debate. Rather than debating only on the ultimate policies, the systems approach enables the debate to be carried down to the very elements out of which policies are formulated: the basic data and the structure of the system can be debated in each part of the system. Of course, in adopting this position the management scientist himself has changed. He is no longer immune to the political power forces of the system, nor can he claim the position of pure objectivity. The scientist now becomes an inseparable part of the system and does not stand apart, so to speak, as though he did not belong to the system.

How should the problem of number assignment for the matrix be resolved in the political debate? *Who* should have the responsibility and authority for making these assignments? The management scientist, who has just given up his political immunity, is essentially in no better a position than anyone else to make the final decision. Nor should it be the head of any of the agencies for they will each have their own political position to maintain. Should we then think of someone who sits on top of the system and can observe every aspect of it in order to make a "fair" decision? Should we say, for example, that it is only the governor's office that is capable of resolving conflicts as to what numbers should appear in the matrix? But to the scientist this would be an irrational way of viewing the problem. The governor's office is also a unit within the total structure of the state and can hardly be thought to adopt a purely objective position with respect to the power politics that goes on between the agencies. Well then, should we say it is the "public" who must decide, because they are after all the ones who are being served by the state government? In other words, the argument would run, if the political debate between the agencies can be examined by the public, then the public can decide what aspect of the politics is truly serving the public interests and what aspect is serving the selfish interests of the agency heads. But here it is unrealistic to expect the public to decide every item that would be inserted in a matrix of the sort described above. Furthermore, this matrix is a fairly small one compared to the type that might arise in other problems of state and federal government.

The management scientist has arrived at a familiar dilemma. On the one hand, he needs data to assess the activities of subprograms, but on the other hand he can't get the data unless he understands thoroughly the whole system. But he can't understand the whole system until he has analyzed the parts; this is the only feasible approach. Ideally he can escape the dilemma by arguing that the political debate is itself a kind of estimating procedure in which considerations of the whole system naturally arise. Thus, the scientist may consider designing a "negotiating component" of the system that is capable of generating the sorts of data that are most suitable in the analysis of problems because it is designed to bring to light the true issues of the total system. As the scientist's systems approach consists of a display of alternative designs and the selection of the best design in terms of the whole system objective, this idea must apply as well to the method by which data are selected by system scientists and the decision makers.

Unfortunately, it would be politically naïve to think that such a negotiating component could be designed and used. In the politics of large systems like state governments, the various components of the system rarely come face to face, and therefore never really understand the reasons that one component feels one way and another feels another way. Indeed, the politics of the state agencies makes it a necessity that each seek its own goal. If a very large figure appears under the Department of Correction in the matrix, then the Department of Correction will construe this as a plot on the part of some agency to reduce the Department of Correction's budget in order that the opposing agency may garner some more "empire" for itself. The scientist nonetheless will do his utmost to get the agencies to meet together, so that each can understand why the other had reasoned as it did, thereby removing the feelings of frustration that occur when communication does not take place.

In any event, however they are collected, the data must be accepted as tentative. For the scientist it will be absolutely essential that in the continuing process of planning and replanning for large systems, he has an opportunity to cycle back through the data base in order to ascertain where misconceptions about the system have led to incorrect data.

To return to the alcoholism mission, so far we have been discussing the activities that currently are going on within the state government that contribute to the performance of the alcoholism mission. But activities by themselves mean nothing. The scientist wants to relate the magnitude of the activities to the total performance of the mission. If he keeps within the existing system, then he wants to determine the *existing* needs and demands for the activities that occur within the state that are related to alcoholism. This analysis requires breaking down the requirements in terms of their origin and constructing another matrix, similar to the one given earlier, in which the entries represent the demands made on the various programs of the alcoholism mission. This can be done by classifying the demands in terms of various kinds of people, the "customers" of the alcoholism mission. The reason for classifying the customers is to evaluate the effectiveness of the mission, as different customers will receive different types of service and gain a different value from the service.

Obviously, the first basic breakdown is between nonalcoholics and alcoholics. Common sense also suggests a separation of youth from adults. Hence we separate the nonalcoholics under twenty from those over twenty. We then subdivide each of these groups into nondrinkers and drinkers. Similarly, among the alcoholics, we might want to break down those who were in the rehabilitative stage into various degrees of rehabilitation, and those who were in the nonrehabilitative stage into various degrees of severity of their alcoholism problem.

Assuming that this breakdown is amenable to data collection, we then construct a matrix (Table 2) which indicates the number of individuals, or, more specifically, the total hours demanded by individuals of each of the programs listed under the alcoholism mission. For example, under the prevention program, we would find a certain demand on the part of nondrinking individuals under twenty, e.g., for educational services, so that these nondrinking youths could be made aware of the problems associated with alcoholism and perhaps never take up the habit. In this case, the scientist could assess *under the existing situation* the amount of activity devoted to educating the nondrinking youth. It is obvious that in the case of this "requirement matrix" the same political problems will arise in the assignment of numbers that occurred in trying to fill in the activities matrix (Table 1). There will be no explicit and objective set of data that provide the accurate set of numbers for the table. Thus the high schools may want to show that there is a real demand for alcoholism education that can only be met by adding more coun-

TABLE 2

PROGRAM	Nonalcoholics				Alcoholics						TOTAL
	UNDER 20		OVER 20		REHABILITATIVE			NONREHABILITATIVE			
	Non-Drinker	Drinker	Non-Drinker	Drinker	Stage 1	Stage 2	Stage 3	Class 1	Class 2	Class 3	
PREVENTION	√	√	√	√							
Educational											
Legal											
Economics											
REMEDIAL											
ACTIVITIES					√	√	√	√	√	√	
Detection											
Screening and Evaluation											
Treatment											
Rehabilitation											
Custody								√	√	√	
CONTROL	√	√	√	√	√	√	√	√	√	√	
Medical											
Economic											
Legal	√	√	√	√	√	√	√	√	√	√	
Social											
RESEARCH	√	√	√	√	√	√	√	√	√	√	
SUPPORT	√	√	√	√	√	√	√	√	√	√	
Total											

selors. Here again the scientist will do his best to create a "negotiating system" for the determination of the requirements. To repeat, however, the politics associated with number assignment in this matrix is a politics associated with the characteristics of the whole system.

The two matrices we have been discussing—the one associated with the amount of activities going on in each subprogram of the alcoholism mission and the other associated with the demands—provide a first estimate of the effectiveness of the entire system. If, for example, one finds that the real demands for some of the programs are quite large compared with those for the other programs and yet the activities as represented in dollar amounts are the reverse, then there is reason to suspect that the present design of activities is not the appropriate one.

If the scientist proceeds to try to set up the budget in terms of existing demands, then he takes the second matrix, which represents the existing demands for the various subprograms, and tries to reconstruct the first matrix so that it more satisfactorily meets these demands. For example, if the greatest demand in the second matrix occurs for first-stage rehabilitation of alcoholism, expressed in terms of man-hours of service, and if this first stage of rehabilitation essentially requires clinical work, then the scientist expects this activity to be adequately reflected in the first matrix in terms of dollars spent, say, in the medical subprogram of rehabilitation. If this does not turn out to be the case, then he believes he has reason to suspect that under the existing conditions the mission is not being carried out satisfactorily and reallocation of funds is required.

It is easy to see how the analysis of the alcoholism mission fits into the logic of Chapter III. Each subprogram is an activity center. The more activity in each center, the higher the overall measure of performance, all other things being equal. For instance, the more preventive education, the less social damage caused by consumption of alcohol. Each activity in principle has a "rate of contribution" to the overall measure, i.e., for each activity there is a coefficient of performance that is roughly estimated from the requirements matrix and the cost of the activity. The optimal design is one that maximizes the overall score by a rational allocation of activities in each subprogram, subject to manpower and funding constraints.

In the actual practice of program planning and budgeting, the requirements matrix is compared with the activities matrix in order to arrive at a judgment about the costs and benefits of each activity. In principle, this judgment is supposed to estimate the coefficients associated with each activity. Recall that in Chapter III each coefficient represented the net contribution, i.e., total benefits minus cost. In arriving at a cost-benefit judgment, considerable effort may be put into the economic analysis of the contribution of each activity. What results, then, is a new matrix in which each activity's benefits can be compared to its costs. Thus the mere size of the activity in the requirements matrix should not be taken as evidence of true benefit.

Suppose, for example, that the school systems begin to initiate courses in alcoholism for the youth; then the requirements table will show a large demand for education in the prevention program for those under twenty. But one might find that these educational programs are of no benefit whatsoever, because the causes that drive the youth into drinking are of such a nature that the mere existence of courses available to them in high schools will mean nothing in terms of their decision either to drink or not to drink. Consequently, the mere size of the requirements from any sector of the citizens of the state does not indicate a true measure of performance, and a more precise analysis of benefit is demanded.

Obviously there are many variations of this scheme of program budgeting. But our concern here is less with the details of the schemes and more with the underlying logic. Why is this approach to budgeting appropriate? The underlying logic, according to the management scientist, must be found in the "purer" allocation models, like the one described in Chapter V. Well, has this logic actually been carried over into the budgeting procedures?

The scientist has many reasons to feel uncomfortable about the program-budgeting techniques, however they are implemented. One of the chief reasons for his concern is that PPB is a technique and to the extent that it is, it fails to be research. Wherever serious gaps in knowledge occur, the budgeting technique must make judgments—often very subjective and ill-substantiated judgments. From the scientist's point of view, the important output of a program-budgeting technique is not the budget itself, but the understanding of the gaps in knowledge.

Once we begin to raise questions about the appropriateness of the existing system, then these gaps in our knowledge become very apparent. For example, it may be true that high-school education is ineffective as a preventer of alcoholism under existing conditions, but a change in the social environment (e.g., the institution of "youth clubs") might greatly increase the benefits of high-school courses. As the scientist sees it, in the consideration of large systems many problems are revealed by means of the systems approach that have not yet been solved, and *therefore* require research programs in order to solve them. The great advantage of the systems approach, says the scientist, is that these problems arise in a very specific and meaningful way, so that we can directly relate the output of the research to the solution of the basic problems of the state government. Thus, if we ask ourselves whether there should be high-school courses and lectures on the effects of alcoholism, we are asking in what environment such courses and lectures could actually lead to prevention of the drinking of alcohol and specifically to the drinking of alcohol that is socially dangerous. It may very well happen that there are past records that provide information about the relationship between high-school educational programs and real prevention, but the analysis of these records can be made only by research methods. If data are not available, then research methods would require the setting up of various kinds of experimental programs.

The point the scientist is making is certainly an important one for social science. As we move from the existing system, i.e., the existing available data, into the deeper problems of whether the existing system is an adequate one to perform a given mission, then it is necessary, he says, to state quite clearly the gaps in our knowledge. These gaps define research programs in social science that are specifically geared to solving total-system problems. Therefore, instead of having a proliferation of social studies dealing with *ad hoc* problems that the social scientist invents, the systems approach provides a plan for social research by showing the particular kinds of problems that need to be solved in order to attain a rational allocation of resources to the missions of the system.

To make the scientist's point clear, look back on what we have done so far in terms of the existing system. As I indicated at the beginning of the chapter, it's quite likely that the drinking of alcohol creates no social disadvantages but instead creates benefits for a class of the drinking public. If we could recognize the characteristics of this class, perhaps in terms of their psychological makeup, then we might eliminate a part of the whole problem by seeing that none of the activities of the alcoholism mission is at all relevant to this class of people. To speculate a bit, suppose we assume that at least half of the human race would never drink dangerously even if given the full opportunity to do so. This half would include people who don't drink by preference, people who only drink very moderately in social groups, and so on. If this is the case, then to apply any of the activities of the alcoholism mission to this half of the population is a waste of time. Their own psychological makeup is such that they are virtually never in a position to cause social damage by their drinking, whatever they drink. If research could reveal the existence of this part of the population, then we would have some basis for considering the real validity of the second matrix. We could ask ourselves what kinds of activities are being directed at the present time toward this "harmless" population and begin to judge whether some of these activities could be redirected to the people who are susceptible to dangerous drinking. The idea behind such a research program is contained in the concept of prevention, i.e., the identification of the potentially dangerous drinker at some early stage at which education, psychological treatment, and perhaps even extensive clinical assistance could keep him from ever arriving at a later stage in his development when the need for alcohol becomes dangerous.

Actually, the scientist had already looked ahead to the argument he just made when he set down the programs and subprograms of the alcoholism mission. There is one item in his list called "research." What he is now saying is that the research program should become an integral part of the systems approach itself and that the amount of research activity should be determined by the analysis of the system. One cannot determine what kinds and what amounts of research are required without looking at the various programs in the detail that the scientist requires. The examination of the existing system points out many deficiencies in our knowledge of how the alcoholism mission works, and therefore leads directly into a definition of the appropriate research program.[1]

You will have long since sensed how the management scientist's arguments have again drawn him, willy-nilly, into the maelstrom of political intrigue and power struggle. Specifically, why should we trust him to tell us how much should go into the research activity, as he is obviously biased? From the point of view of the whole system, the question is how much of the available funds should be spent in research in order to improve on the existing system *versus* prevention, or rehabilitation. To answer this question, we must estimate how much research would improve either the prevention or rehabilitation programs. But even the scientist has to admit that research always takes place in a state of ignorance. Perhaps, to know exactly how much a given research effort would contribute to a given program, we need to have conducted the research.

[1] It is interesting to note that for many administrators of alcoholism programs, "research" means medical research, and not social research of the type discussed above. Thus the need for research often does not fall out of a systems analysis, but rather out of a "felt need."

Furthermore, the state's policy with respect to research in alcoholism cannot be separated from its policy with respect to research in various other aspects of society that interest the state government: education, health, law, etc. This means that the alcoholism mission overlaps with the research mission of the state. This is as it should be, because the overlapping highlights the question of whether the existing system of state activities is an appropriate one. This is even a broader question than whether the existing alcoholism mission is appropriate.

The management scientist is apt to find himself swept out to sea by this stream of questioning, thus finding himself far from the dry land of feasibility. His thinking has led him to wonder about the social benefits of research and development activities with respect to all the various missions that a state government carries on.

At the present time the management scientist has no technical means to assess the social benefits of research activities, even though the issue is one of critical importance. Thus in the United States, the federal allocation of research funds may be entirely wrong. About 90 percent of the federal government's expenditures in research and development go to the military, the space program, and to the nuclear-research program. Although all three of these areas are important, they are all devoted to the development of various kinds of technology. It's true that the Department of Defense has pioneered in the development of program planning and budgeting that we have discussed in this chapter, but its program planning and budgeting is all oriented to the defense mission. Consequently, at best, the Defense Department has been able to look at the existing defense mission and ask whether within the budgetary constraints it can be improved. It has not asked whether the research activities supported by the Defense Department are appropriate in view of the demands for research activities in the nonmilitary sectors of our society.

There is certainly good reason for skepticism on the part of many politicians and industrial managers concerning the usefulness of research on social problems. The skepticism has arisen out of the fact that so much of social research is conducted in a fragmented way in which enormous amounts of data are collected, correlated, and filed away in reports that at best have a mild interest to the reader, and at worst are totally irrelevant for decision-making purposes.

Although the management scientist may feel that the systems approach to program planning and budgeting may be the answer to the haphazard proliferation of social research, he is in no sound position politically to implement his ideas. On the one side, the academic community does not want to be told what research to conduct; it still espouses the traditional idea that excellent basic research is born out of an intellectual curiosity. This curiosity itself is created within each individual—or, more likely, within each "college" of academically minded individuals in the various disciplines. On the other side are those who disperse funds; what compelling arguments can the scientist develop that will convince those who control funds to support research on the "gaps" that the scientist finds in program budgeting? Whatever arguments the scientist uses cannot be based on one of his "models," because he doesn't know how to model the benefits and costs of research. But even if he claimed to have such a model, why should the decision makers trust him? It would be much as though a labor union developed a model that "proved" higher wages and fringe benefits were socially desirable.

Thus, the management scientist finds himself in the awkward position of being neither totally pure nor totally impure. He is not a pure scientist because he takes the source of his research from the gaps he finds in trying to implement a systems approach, and not from his own intellectual curiosity. On the other hand, he is not totally impure, either, because he is reluctant to become a politician, as it is not feasible for him to analyze the power struggles of the whole system.

These concerns of the management scientist have much to do with the definition of the "systems approach." On the face of it, the systems approach means looking at each component part in terms of the role it plays in the larger system; but there appears to be a deep paradox in this rather obvious prescription of rationality. The paradox is of the "Who shall decide?" variety. Who shall decide how to look at each component, when "looking at" means "conducting extensive research"?

Paradox is not the end but the beginning of the formulation of an idea. It suggests where to go next. In this case, the next step of the scientist seems almost obvious. He has been brought to the point where he sees gaps in the information he believes is required to make good decisions, i.e., decisions based on a systems approach. This suggests that he had better delve more deeply into the role of information in the managing of systems. Perhaps if he can better understand the proper design of management information systems, he will be in a sound position to evaluate the benefits of additional information and hence be on his way to a sound argument for more research.

7. MANAGEMENT INFORMATION SYSTEMS

In many of the discussions of the systems approach one finds the concept identified with the utilization of computers for processing information. A company or a government agency may feel that it has taken a systems approach to its problems if it has asked a computer company to examine its information system and determine how the computers can be used.

Of course, by now we know enough about the management scientist to predict that he will not say that computerized information processing by itself constitutes a systems approach. Now computers are undoubtedly a system; hence those who design computer hardware and software programs have to think about the entire computer complex from the systems point of view. But even though the computer department of a company is well designed from a systems viewpoint, this does not mean that its activities in the company constitute a systems approach to the company's problems. Just because certain functions in a hospital are connected together by an information system that is handled in part by a computer does not mean that the hospital has taken a systems approach to its problems.

But as the management scientist wants to understand the role of information in the management of systems, he will certainly have to consider existing and past information systems and will expect to find many clues in the designs of computerized systems.

Among the largest information systems we have inherited from the past are the traditional libraries. It will help us to understand how an information system works within the larger system if we examine some of the traditional policies and designs of library systems. One aim of the traditional library is to collect documents of various kinds, books, articles, maps, etc., and so to identify the collection that an individual who wishes an item from it can retrieve it "successfully." The measure of performance of the system may be given in terms of the size of the collection and the ability on the part of a user to retrieve a document from the collection. These two measures can be reduced to one, the probability that any user of the system obtains the document or piece of information that he wants. But from the systems point of view, this measure by itself is not an appropriate one because it would imply that the larger the collection the better the library, an implication that certainly is not true in general.

However, suppose we proceed as though the measure had at least some merit and examine how such an information system works. The management scientist, as always, will set out to identify the components (programs) of the library. These might be: solicitation of documents, reception, cataloguing (identifying), storing, retrieving, advertising, and research. If we look at reception and cataloguing in the traditional library, we see that most documents that are sent to the system are not rejected. The point here is that, as the system cannot be aware of all possible requests, it should store an item unless there is a strong probability that the item is worthless or never will be requested.

The traditional system, of course, is far from being passive. It attempts to connect together the documents stored in the system in various ways, as the user quickly finds when he consults the catalogue. Perhaps a convenient term for this activity is "construction of fact nets," i.e., the interrelating of various pieces of information in the total system. The fact nets enable the user to retrieve a document and to be informed about the existence of other documents that might be of interest to him. Once he attains the additional documents, further clues may lead him to other pieces of information that he needs.

The normal activities of receiving, cataloguing, and constructing fact nets are supplemented by various kinds of special staff work in the traditional library. This work arises from specific requests made by users for the assembling of information in a certain form, e.g., assembling of demographic information, or, in the case of industrial companies, assembling of information about past sales or past costs.

Would the scientist feel satisfied in saying that the traditional library is well designed in terms of a systems approach? This question, as we've seen in the previous chapter, is somewhat ambiguous. As we indicated there, one stage of the systems approach is to look at the system as embedded in an existing system and its actual requirements. Another way to look at the system is to raise questions about what the system would be like if the larger system in which it was embedded were radically changed. But even when the scientist looks at the traditional library within the existing system, he has a number of reasons for being skeptical about calling its design a "systems design."

First, it seems obvious that the programs of the library are not being examined in terms of missions performed throughout the larger system in which the library is embedded. Where else in the system, for example, is the acquisition of documents performed?

In a university, the programs of solicitation and reception of documents may be conducted in a number of places within the university other than the library. The individual departments and research institutes normally solicit and receive an enormous number of documents, as do individual professors. The same point can be made about government-agency libraries and libraries of industrial firms. The collection of items of information goes on in many places other than the officially recognized central information system of the library. If the librarian were to reply to the implied criticism by saying that, except for the library, all the other document collections are for very special purposes and very special use, then the scientist would insist on an analysis of the demand on the various activities of the program, just as in the last chapter he analyzed the demand on activities that were carried out in sections of the state other than that which was directly responsible for the alcoholism mission.

Second, the scientist finds that the programs of the traditional library are not related to the measure of performance of the library. The measure of performance is in terms of the benefit to the user, minus costs. Now, many users come to the library with a specific request which they try to match against the catalog or the memory of the reference librarian. A very significant part of each request, namely, that the retrieved information be reliable and truly relevant to user needs, cannot be satisfied by the traditional library. In fact, the traditional library has to change the nature of the user's request for knowledge into a request for a specific set of documents; the library does not have the responsibility for guaranteeing either the validity or the relevance of the documents.

Finally, the traditional library is organized into various departments, and costs are determined along departmental lines. Hence, the costs of the various system activities are not related to the true requirements for these activities.

Thus the scientist would find it extremely difficult at the present time to study the library of a university and try to construct a model of its effectiveness. He would not be able to tell whether the activities of the library are contributing to the measure of performance of the library system, i.e., to user benefits. So he would think it quite impossible to determine whether the allocations that are made to various programs and subprograms are proper. And lastly, it would not be possible to determine the real value of the library for the larger system in which the library is embedded.

These conclusions apply to any of the well-known library systems, the Library of Congress, university libraries, community libraries, the libraries of industrial firms, libraries of research organizations. Consider, for example, the community library, where citizens may come and secure books for two weeks "for free." Although it may seem to the average citizen that he is obtaining free reading by this means, if he were to estimate the cost to the city of buying and retrieving a book he borrows, he might conclude that it would be far cheaper all around if people bought their books in paperback bookstores. The measure of benefit to the community of the community library is usually obscured by a number of unanalyzed factors associated with the appropriate amount of activity that should be carried on in each of the library's programs. For example, there may be a feeling that the turnover of books is a good measure of performance of the library; i.e., the more frequently a book is taken out, the greater its utility to the system, and the greater turnover of books there is in the total community library, the better the library's performance. But turnover *per se* is merely a physical measure of the activity of retrieval for users.

As the scientist sees it from the systems point of view, one cannot use the amount of physical activity as a measure of performance of a system. One has to show how the activity is translated into a measure of utility or value. For example, one way to get a great deal of turnover in a community library is to advertise and store pornographic literature, or literature that essentially tells harrowing stories about Communists, Nazis, or the underprivileged. In these cases, is it appropriate to argue that the turnover constitutes a true measure of the performance of the library? Perhaps the retrieval of books on health or education is a far more significant type of activity than the retrieval of popular novels or books on travel, and perhaps not. The point is, until we understand the relationship between the activity and the utility of that activity to the system, we are in no position to view the community library from the systems point of view.

Now it should be clear why merely computerizing the traditional library as it has been described above does not constitute a systems approach for the management scientist. For instance, there is some interest today in putting the Library of Congress catalog into computer memory so that the user can interact with the computer when he requires information. But this development does not constitute a systems approach to the library for the scientist. It does not because the basic systems questions associated with the catalog and its usage have not been answered merely by putting the catalog into machine-readable form.

The example of the alcoholism mission of the last chapter indicates how the scientist would expect to move toward the systems approach to libraries and other information systems.

First of all, in order to include the needs of the user, the scientist would want to design what might be called the "quality filter." This would be a program which is directed explicitly at filtering unreliable, irrelevant, and meaningless data, by means of some kind of expert judgment, at the point when the information is received, and later discarding information when it is no longer of any use.

But this is not all that the filter should accomplish. Even though it is relevant and reliable, some information should nonetheless be "forgotten" by the information system. This is more or less obviously true if the relevance of the information does not pay for the cost of its recording, storing, and retrieving. But there may be a deeper role for the "forgetting" function in information systems. Every human being knows the advantages of forgiving and forgetting; they form a part of his living relationships with others. However, when we come to large organizations, we seem to neglect the lessons that individual living has long since taught us. We feel that, just because we can record and store information that is relevant, we should do so, "in case at some later period of time we would like to recall it." But the very having of the information may turn out to be a severe disadvantage in later activities of the large system. The profession of law has long since learned the great advantage of forgetting information. If all information were used within the legal profession, the profession would be bogged down in a helpless mire of legal confusion.

An excellent example of the evils of the failure to forget lies in the FBI and similar files. The principle behind the files seems to be that information should be collected about every individual of the United States if possible so that "later on," in case his name comes up for security clearance or criminal action, one can retrieve all relevant information about him. The criterion of relevance is quite broad, e.g., anything said by anybody else about the person may be regarded to be *prima facie* evidence of the relevance of the information.

This means that, when the individual is examined later on—for clearance, say—then all items of information have to be retrieved and explained as the information storage itself has permitted most "information" to get through its filter. Hence the file never forgets, and because it never forgets, the types of evil such files create throughout society may often be far greater than the claimed benefits. For the scientist, there is a need to adapt the systems approach to such files by introducing a measure of their effecitveness for the citizens.

But in trying to adjust information systems to user needs, the management scientist is bound to run into opposition. First, there is the efficiency-minded person, who will argue that the failure to keep information that later on would have proved useful is an inexcusable error. Furthermore, he would argue, the development of the technology of information storage, on microfilms, tapes, and disks, yearly reduces the cost of storage and retrieval. If the management scientist now argues that the problem needs to be analyzed by a model that balances the cost of storage with the cost of failing to have the information, he immediately runs into his old problem of the infeasible. How can he predict whether information will be useful or useless? The problem is not too different from the problem that automobile manufacturers face in keeping spare parts; some owner of a 1925 model is bound to show up sometime. Similarly, even though a document hasn't been requested for years, someone is bound to come along and request it, unless its uselessness is obvious. As for the immorality of remembering, that's a problem for others to solve, not the information system. No library or file system, runs the argument, should have the responsibility of determining how information is used; how could the information system control user behavior?

The scientist can certainly sense the reasonableness of these arguments; he is also aware that he lives in a magpie civilization that tends to save everything it can: buildings, books, and bureaus. His obvious tactic is to call for research on information usage, but he can easily sense the opposition that this call will generate. Information *theory* is a safe science, as it deals with the transmission of messages and not their meaningfulness to the user. As soon as we get into research on user behavior, we turn into snoops, people who pry into the affairs of other people. Not only do the humanists object, but so do the technologists, who claim that the research is not apt to reveal anything worthwhile.

Thus the scientist finds it almost impossible in the case of information systems to assess the value of research on the behavior of the user. If he is technically trained, he will probably turn to the safer topics of research: the structure of files, retrieval techniques, automatic abstracting, and the like. His approach will consist of assuming that the larger system in which the information system is embedded is a satisfactory system. That is, he will try to design an information system assuming that the existing system is proper, just as in the last chapter he began by designing an alcoholism mission assuming that the existing state system was proper. He will therefore assume that an information system is performing satisfactorily if it "satisfies" the requests that are made, even though the explicit requests may not represent the true needs.

Thus most computer-based information systems today operate according to the traditional attitude toward demands: the information system itself has no direct concern with the legitimacy of the demands. Its measure of performance lies solely in the satisfaction of demands, however they are made. The systems are designed to tell the user what is in the document, and they screen for quality in a very broad sense, but beyond this, they do not go: *caveat emptor.*

Nevertheless, the debate need not stop where present practice does. Despite the fact that great claims are made for computer-based information systems, and these are sometimes portrayed as *the* systems approach, in a very fundamental sense the scientist believes they fail to represent his philosophy of systems. They fail simply because their measure of performance is in terms of the transaction, rather than the benefit. The true benefit of an information system must be measured in terms of the meaning of information for the user.

Suppose, then, in pursuing this idea, we explore the notion of meaning, even though the possibility of empirically testing it may be some distance off. As the scientist enters into these speculations, he passes beyond the existing system and tries to imagine how a design might be accomplished which would improve performance in some real sense.

First of all, in a typical scientific gesture, he makes a distinction. This distinction rests on his observation that in many cases an individual needs information in order to fill out a picture that is almost but not quite complete. The specific piece of information needed is, say, one item that will complete the picture; once he has the item, the user can decide correctly.

Imagine, for example, a man in a hotel when the fire alarm has rung. As he rushes down the hallway, the question in his mind is which door leads to a safe exit. Note in this case that practically all of the decision making already has taken place; but his "model" of the situation lacks one item, the correct exit. Once that fact has been retrieved, then the correct decision will be made. If the man running down the hall has to open various doors to determine which is an exit and which is not, then the cost of retrieving the fact may be disastrous, whereas if there is a sign saying "Fire Escape," the information system has generated information which, from the point of view of the escaper, has an enormous value compared to its cost.

Instances of this type are called "fact retrievals." In such cases, we do seem to have, in principle at least, a way of identifying the true need. The value of the fact retrieval is the improvement in the user's behavior minus the cost of gaining the information, given that most of the basic policy of the decision maker has been settled beforehand. For instance, an engineer who is trying to design a bridge in a certain manner may need to know the strength of certain materials. He will go to a handbook and retrieve the "facts." Note again that the basic policies and the underlying model have been decided upon. The engineer is not speculating about whether a bridge ought to be built or even at this stage what kind of a bridge. All such deliberation has gone on ahead of time. In the same way, a traveler who wonders how to get from San Francisco to Asheville, North Carolina, will go to the airlines manual in order to find out how to make various connections. He will be engaged in fact retrieval, because the basic policy—i.e., the advisability of the trip and the mode of transportation—have all been settled ahead of time.

Even the scientist who is strongly oriented toward feasibility and hardware design sees how to design and evaluate a management information system that is related to fact retrieval. If the user population is sufficiently well defined, the systems designer need only determine the probability that a fact of a given kind will be required and what its probable value will be. He then wishes to minimize the cost of retrieving a fact when it is needed, in terms of both money and time. This is the principle by which airlines schedules, many accounting systems, engineering handbooks, and, today, some computerized systems are designed. Given the evolution of the computer, we can expect to find far more sophisticated fact-retrieval systems in the future. The construction engineer, for example, may be able to purchase a console on which he can type in his request for information, and in a "time sharing" mode receive an answer in a matter of seconds.

But note that even in the case of fact retrieval the designer of the system has to perform some very difficult tasks with respect to the user. Not only does he have to determine the value of the information by some means, but he also has to store in the system the model or policy of the user. For example, the airline schedule implicitly stores a model of the user in which the assumption is made that the air trip is essential. The guide does not question the user, as the government did during World War II, about the advisability of the trip in the first place. It does not say: "If you were to employ a telephone or even a letter, you would accomplish your purposes far better and at far less cost." And yet, this statement itself must be regarded as information. In other words, the designer of a very simple fact-retrieval system may be failing to supply the most relevant kinds of information when requests are made. The "fact" that is needed may not be the fact regarding the departure of an airplane, but the fact of the availability of another communication device. Indeed, there is the serious question whether highly "efficient" fact-retrieval systems may not tend to stifle creative questions, e.g., questions about the user's whole system. Again, pure efficiency of operation may turn out to be very costly.

Then, too, there is the question of the reliability of the fact. An airlines guide is a forecasting device. Most travelers are not interested in how planes traveled yesterday; they are interested in how they will travel tomorrow. But the airline guides do not indicate that there is a certain probability that the planes will not depart because of weather conditions or machine failure. Yet this is a very relevant piece of information from the traveler's point of view. The reliability of a "fact" in the information system depends on a forecasting model. In some cases this model may have a high reliability, because it is well embedded in traditional disciplines of science, whereas in other cases it may not.

We can see now that the systems approach to management-information systems implies an extremely important and yet difficult problem: What kind of model of the user should be stored in the system, and what should be the reliability of the model with respect to the facts?

Consider the richness of the fact-retrieval system that is stored in the head of an expert compared with the information system that is stored in a library. The library system obviously has many more "facts" than does the brain of any particular individual, no matter how much of an expert he is. And yet if a scientist were to turn to an expert friend of his and say, for example, that he is bothered by a certain problem in the field of communication, his friend may very well reply, "Well then, you should read Jones's latest book on the behavior of bees." His friend can do this because he has a very clear model of what the scientist is trying to do and hence can relate the real intentions of the scientist to work going on in totally different fields. But in the case of a library catalogue, the request has to be taken literally. The library will use its cataloguing techniques to track down relevant facts once they have been identified by various terms. The library therefore has a minimal model of what the user is like.

There is one final point that needs to be made about so-called fact-retrieval systems, a point to which I alluded earlier. This is the issue of the illegitimate use of facts, in whatever form they may occur. The files of the Internal Revenue Service are not open to every user, nor are medical files generally available, because in both cases the risk of illegitimate use is too great. Therefore, it is simply not the case that information systems have no concern about the user's intent. As the scientist sees it, we must build into the design the safeguards of privacy and, more generally, the safeguards against illegitimate usage. But this is no easy matter to decide on. For example, should all educational records be open to any citizen willing to pay the price of retrieving them? Remember that in many instances these records contain confidential letters written about the student.

As the scientist in his speculation moves to richer information systems, in which the model of the user's decision making is far from being complete, he will have to consider more and more deeply the kinds of presuppositions that need to be stored in these systems about the users' needs and resources. The models that need to be stored are representations of the user of the system. The representation depicts the kinds of things the user is trying to do and can develop forecasts of what will happen if he adopts certain policies.

Consider, for example, the industrial manager who is trying to formulate some plans. It may soon be possible, under a "time sharing" system, for the manager to have a console and other display devices in his office. If he wishes to speculate about a new product or an advertising campaign, he can query the central computing system. The system will not be constrained to fact retrieval. In addition, it will take the questions that are being asked and frame them in a mathematical model of the manager's problem. It will then develop forecasts that compare the policy he is considering with alternative policies and will display the results to him. The manager can then interact by asking additional questions or modifying his policy. The additional questions, or additional information from the user, will be sent via the office console to the central system, which will incorporate them and, when necessary, modify the model or the data bank. In effect, such an information system would be simulating the kind of rich interchange that occurs between a scientist and his peers, or between a manager and those who are well informed about the manager's problems.

The following little anecdote illustrates the way in which such an information system might work in the scientific community.

Loneliness

He opens the door and steps into his study. His desk is already well laid out. The manuscript he has been working on is ready, with an adequate supply of pencils, pens, and scrap paper. On one side are the pages he worked on yesterday, returned from the Computer Center with its queries, notations, and additions.

On the other, neatly stacked in piles, are the several kinds of information he will need for today's work. All this annoys him, as it has before, because there is no possible excuse for delaying action as there used to be in the old and inadequate days: sharpening pencils, finding paper, browsing a bit, getting coffee. The coffee is already steaming cheerfully beneath the lamp-and-urn that was set off when he opened the door.

Reluctantly but of necessity, he sits down and begins to go over yesterday's manuscript. This was picked up by the Center after he left yesterday. The Center's information system has reviewed it as follows:

(*a*) It has edited the writing for grammar and spelling.

(*b*) It has noted obvious stylistic weaknesses: repetitions of ideas, awkwardly phrased sentences, etc. The Center has printed out suggested new sentences or parts of sentences, in such a manner that he need only cross out the old, or the new, as he wishes.

(*c*) It has scanned the writing for its correctness. Each declarative sentence is examined for its contents and ranked in terms of the importance of the content. The most important sentences have been compared with the Center's "relevant memory," which is essentially the storehouse of what has been written in his field. In each case the Center scores the sentence as "redundant," or "opposite to agreement," or "new, but unopposed."

(*d*) The Center has abstracted what he has written and compared this abstract with its total memory of ideas. The output is simply a series of suggestions as to what might be relevant in some other fields.

(*e*) The Center has supplied a series of references for the material submitted.

Because it is important that our hero not spend all his day reviewing what the Center had to tell him, the Center has prepared a list of the items, ranked in terms of importance, so that if he chooses he can ignore some of its responses. Thus it might say:

1. Read Ageldorf's *Möglichkeit und Erfahrung.*
2. Paragraphs 6 and 12 partially contradict each other.
3. Paragraph 13 has no significant sentences.
4. Paragraph 20 says the same thing as paragraph 4 of yesterday's manuscript. [The Center, of course, stores all previous manuscripts and its replies.]
5. Sentence 3 in paragraph 25 is incomprehensible; sentence 5 is a tautology.
6. You are spelling "tolerance" incorrectly throughout. [Spelling errors always come last, of course.]

The first file of documents on his right consists of items that the Center thinks he might like to read. These are abstracted and ranked in terms of their importance. The second file is an answer to his queries of the day before, which he had sent via his typewriter. Some of these queries are specific questions about historical fact, some are for references, some for "who else worked on this idea." One of them is a mathematical problem. He had been working on a theory that large organizations develop a catatonic state under certain circumstances, and he wished to describe some of this process by means of the language of "trapped states." The Center had worked out a theory along the lines of his suggestions.

Finally, there is a general pile of material the Center thinks he should examine "at his leisure." One of them is a recently published detective story. The Center has made no evaluation of his work. This will come later, when the Center's reviewing system will tell him whether the work is publishable, where to publish it, how to condense it, and so on. Sometimes the Center will interrupt in the middle of the work and query an author whether, in view of the large number of redundancies, tautological sentences, and contradictions, he wants to continue. The Center never recommends resignation or other measures. (The Center has been known to exclaim "Wonderful" about some manuscripts, but proper debugging has fixed all that long since.)

As he works he may query the Center. If he pushes a red button, he wants a reply as soon as possible; a yellow button says "at the end of the day," and a blue one has "mañana" printed on it. His queries can be for references and abstracts, for the newness of an idea, for the clarity of an idea, for the solution of an analytical problem, etc. It is important that on occasion the Center reply immediately, because that's the way the two of them get to know each other.

In the early days, there was a good deal of give and take, in which he would tell the Center to quit referring him to Harbison, "who is an ass," or to try again because the last reply was irrelevant. By now, however, the Center has come to know his ways, and they operate very smoothly together. The Center has its own sense of humor and will occasionally tell him he's getting too pompous. He likes being told this, though he won't admit it to himself. But he never tells the Center to stop the nonsense, as he has when the Center accused him of currying favor with the President. He also kids the Center for its anti-labor policy, which pleases it very much.

A buzzer sounds on his desk. A voice that is remarkably human informs him that this is the dean with his monthly rating. His activities have been scored for content, for the number of references made to them by others, for their "praise" index, for the number of times his name was mentioned in faculty clubs (all of which are wired to the various university centers). He can see that his "Reference Associates" have fallen down on the job, and he plans to retaliate by cutting a few of them off his list. In general, things haven't gone well. His fame index is lower, his prestige index has sunk below the "red line," and the probability of an honorary degree has sunk 12 points. Discouraged, he decides to quit for the day and enjoy himself. He pushes another of his many buttons. A voice that is surprisingly feminine answers. . . .

It will be seen from this anecdote that in addition to the technologically pleasing there will be some real dangers in future information systems. These dangers lie in the larger world in which both the user and the information system are embedded. About this larger world—in the story, the political world of the academic community—neither the user nor the information system knows very much. More precisely, both the information system and the user have an implicit assumption about the larger world in which they are operating, e.g., that science is pure and politics will never interfere. If this is the simpleminded assumption of the professor and his information system, then the issue about the purity of the world, so to speak, will never come to the attention of either until suddenly there is a jolt from outside that will destroy both the user and his system. In other words, the same old theme has come back to plague the management scientist: he cannot properly separate the subsystem from the larger political system. Indeed, he may make things worse, rather than better, because the ease of use of the information system may blind the user to other stimuli. The user makes very basic assumptions about the nature of the larger system in which we live, which he himself is not even aware of. The reports from the information systems will be fitted into his basic assumptions in such a way that he can never learn anything new about his idea of what the world is like. Everyone recognizes this point when he tries to discuss problems of religion or politics, because we carry on such discussions in closed communities. The information circles around within the closed community and the basic assumption about what the whole world is like goes unchallenged. A dyed-in-the-wool Republican is not apt to be disturbed in the least by anything that he reads in the magazines and papers because the magazines and papers he reads are all Republican-oriented. Even when he sees a radical-left paper, he considers it to be silly; his information system automatically rejects the reliability of the information contained in such documents.

The humanist will detect a great danger both in current information systems and in those which the scientist has been speculating about. It is a danger that makes the humanist question the whole "systems approach" in any of its forms; it is the danger of closure, i.e., the inability of the information system and its user to have proper sensing devices with respect to the outside world. Specifically, the user and his information system may very well be cut off from the deeper feelings and realities of people.

Assuming that this "bleeding hearts" complaint of the humanist makes some sense, how might the scientist reply? Could he design an information system that has some capability of listening to what is going on outside in the larger system?

The Germans have a nice word for the underlying assumptions about the nature of the world. It is a picture of the world, a *Weltanschauung*. What is being asked is how the user and his information system can become more and more aware of the nature of their common *Weltanschauung* and thus raise questions about its correctness.

Imagine, for example, a decision maker who is considering the wisdom of a certain policy he has been following. To be more specific, consider again the policy of the United States with respect to its expenditures in research and development. The policy the United States has been following is to spend about 90 percent of its research and development funds of some 16 billion dollars in military, space, and nuclear energy. Is this a sound policy?

Now, if we had the type of information system the scientist was speculating about earlier, with a built-in economic model, the policy maker would go to a set of data and raise certain questions about the consequences of the United States' research and development policy. The management-information system would forecast the consequences in such a way that the decision maker could judge whether the present policy was a sound one. For example, the system would help him judge whether or not the present expenditures are really essential in terms of the demands that are being made on the defense capabilities of the United States and would also consider the possible utilization of these funds for research in the nonmilitary sector, e.g., in health, education, and the feeding of the world's population.

But all along in this interaction between the decision maker and his information system, there would be a number of underlying assumptions that are never questioned. For instance, both the decision maker and the information system would assume that the United States *should* play a leading role with respect to both defense of "the free nations" and their economic development. But perhaps this is not so; perhaps as a culture we are too immature to take on world leadership. Both the decision maker and his information system would assume that the United States has a responsibility to carry on domestic programs of poverty, transportation, education, and the like. But perhaps this too is not so; perhaps the federal government should stop trying to be an indulgent parent.

Now suppose that the information system were designed in the following manner: When the decision maker poses a question concerning the advisability of current policy, the information system generates an opposing proposal, a "deadly enemy" proposal. In this case, the deadly-enemy proposal might be that 90 percent of total expenditures of the United States on research and development should be devoted to nonmilitary, nonspace, nonnuclear-energy activities. In doing this, the information system would present to the decision maker another picture of what the world is like, a *Weltanschauung* that is different from his own. This second *Weltanschauung* might assume that the world is essentially a world of economic forces and that these forces can be brought under control, e.g., by feeding and educating the world's population. When this happens, says the *Weltanschauung,* the dangers of aggression will disappear. Thus there will be two world pictures operating. The first world picture depicts a decade or more of aggressor activity around the world, in which the United States must keep up its military capabilities in order to police cold and hot wars wherever they occur. Specifically, it must suppress "unsatisfactory" revolutions. The second world picture shows the world to be basically economic in structure and argues that the United States can best succeed where it can change the economic conditions so that revolution cannot thrive.

Both world views in the management-information system we are envisaging can take the set of "facts" stored in the information system and interpret them in defense of the policy they are arguing for. Thus the defense world view will take information about the buildup of missiles in aggressor nations as evidence for the current policy. But the nondefense world view will take the buildup of missiles as evidence of a fear of the militaristic intentions of the United States, a fear that could be overcome if we spent more of our research and development money on the problems of negotiation between conflicting parties, for example—more in the State Department than in the Department of Defense.

The idea behind this "dialectical" information system is to make clear to the user the basic assumptions that go into the support of any proposal. By becoming self-conscious about his assumptions, the manager is supposed to become a better decision maker, for his sensitivity to the world is increased. This might happen when the manager recognizes some common underlying assumption on both sides of the debate. For instance, in the case in point both proposals have a ground of agreement as well as a ground of disagreement. Both assume that the current expenditures in research and development are appropriate; they are debating only the allocation of these expenditures. Both world views assume that the United States is the nation that should be taking the active and leading role in world politics. Neither considers the possibility of the United States' supporting the research and development of the United Nations in defense as well as nondefense activities.

It is by the technique of revealing the underlying assumptions of the data themselves that this speculative information system is able to highlight the importance of the user's view of the world. At the beginning of this chapter the scientist assumed that items of information have a dominant role in the management-information system, so that each piece of information has an existence of its own, and the user is in the position of either wanting the information or not wanting it. As he moved toward a systems approach, the scientist recognized that the user needs to know about the reliability of the information and subsequently that he needs to "insert" the information in a model. What was not clear in the earlier design of an information system was that an item of information carries with it an implicit view of what the world is like, specifically, that the utilization of the information makes certain very strong and often implicit assumptions about the larger system in which the user and the information system are embedded.

It is in this area that research on information systems might unearth extremely useful and fruitful ways in which the user and the information system can interact. When we consider that most of us as citizens are faced with an enormous number of problems that we must try to solve, both in our community living and in city, state, and federal governments, then it begins to look as though it were a hopeless task to keep the citizen well informed. The same remark, of course, applies to people we elect as our representatives in legislative bodies, as well as to the managers of industrial firms, educational institutions, hospitals, and so on. In every case there are far too many problems for the decision maker to try to solve in depth. But, says the scientist, it may be possible to develop a "dialectical" technology whereby the decision maker in the limited time available to him is exposed to forceful debate about the issues that confront him, not a debate between people but one between ideas.

Thus the scientist. You can see the dehumanizing theme up to the very end of the dicussion. The debate, he says, is not between men but between ideas. He acts as though he believed that people are information-processing machines. Indeed, in one area of scientific research, called "artificial intelligence," it is clearly assumed that intelligence is a type of information processing, and hence that computers can think because we can get them to simulate the information processing of people. It's strange how often the critics of artificial intelligence object to the wrong thing here; they are horrified at the suggestion that computers can think, whereas they should be horrified at the suggestion that people are information processors. In a person with deep feelings, there must be a strong opposition to the idea that the design of social systems should be put in the hands of people who don't understand people. In his own terms the scientist fails to understand the whole system. People are not "better" just because they process information more rapidly or more coherently. People are better because they are better in a moral or aesthetic sense.

Oh, says the scientist, I see what you mean. You're talking about values. But I can handle that area quite well, as we shall see later on.

8. AN ILLUSTRATION

In view of the large amount of speculation that went into the discussion of management information systems, it may help to look at one specific example. An appropriate example occurred in the early 1960s when Governor Brown of California issued an invitation to the aerospace firms of the state to respond with proposals for a "systems approach" to some inportant social problems. The governor's idea was that the aerospace industry was well populated with systems scientists who seemed to be doing a marvelous job in the design of systems of various kinds, for the National Aeronautics and Space Administration, the aircraft industry, and the Department of Defense. His thought was that the "know-how" of these scientists could be turned to the considerations of critical problems of the state, because the know-how consisted essentially of a systems approach to problems. It's not clear from the governor's speech or from the various editorial remarks made in the papers whether anyone was too well acquainted with what the systems approach meant, other than it was the approach used by existing system scientists at work on various kinds of hardware systems.

The invitation was for proposals on the problems of information, transportation, crime, and sanitation. It will be of some interest here to look at the information system that was proposed as a result of the governor's invitation, and to examine the design under the five aspects of the systems approach given in Chapter III.

First there is the objective of the state information system. It's easy to formulate the "propaganda" version of this: "The purpose of an information system in California is to provide the public and the managers of the various agencies with the right kind of information at the right time and with the right precision and in the right form as the needs require." Of course, this statement of the problem is merely a way of saying that for any information system you want the very best. The system designers translated the propaganda version of the objective into a more feasible and practical version: to provide information to the public and to the agencies of the type that now exists and is now transmitted by one agency to another or by an agency to the public or by the public to the agency, but to provide this information in a computer-based system at least within the time of the present "man" system, and at least within the costs of the present system. Consequently, the measure of performance of the new system was described in terms of how much improvement the new system makes over the old one in terms of time and cost.

Judging by the discussion in the last chapter, we can easily question whether this measure of performance and its related objective are truly legitimate objectives of a state information system. Information, as we said, has a tendency to accumulate and the more information accumulates, the more information is needed in order to keep track of the accumulation. Information in effect is a reproductive organism that has no morals and goes around generating offspring without any consideration of the effect of its own "population explosion." So, to design a new system that simply tries to beat the standards of the old may be merely to design into fast-moving hardware some of the evils of information collection as well as its benefits.

But the thoughts of the last chapter also implied that going too far beyond present requirements for information is not feasible. Indeed, the governor's invitation did not include an invitation to look into the management and policies of each state agency. Hence the system scientists undoubtedly felt that their mission was constrained to a design of an information system to serve existing needs.

The environment of the information system includes the budgetary limits set by the state government for the acquisition of new equipment and personnel. Given the constraints mentioned above, the environment also consists of the actual information transmittal between agencies and between the public and the agencies. The system scientists therefore developed a large "matrix" to describe this environment. The matrix shows the amount of information that passes from one sector of the state to another sector of the state and is much like the requirements matrix of the alcoholism mission. For example, an immense amount of information passes yearly from the public to the Department of Motor Vehicles and from the Department of Motor Vehicles to the public, just as large amounts of information pass from certain sectors of the public into the educational system and back to the public again. The matrix indicates the requirements that must be put on various information units within the agencies.

The resources of the information system are defined in terms of the allowable budget and the available personnel who would be willing to work as programmers, file clerks, and the like. Resources also consist of existing files. In terms of new technology the resources include the kinds of computer-data banks that are sold on the market—tapes, disks, and drums; new computer processors; and computer links. A very important part of the systems approach to a state information system consists of a careful evaluation of existing computer hardware in order to determine which items on the market are most suitable within the budget allowed by the state. This study of resource allocation is conducted in terms of cost-benefit analysis, usually by comparing two or more specific pieces of equipment. Sometimes the comparisons require rather subtle judgments about "trade-offs," e.g., the trade-off of processing time *vs.* cost, or programming costs *vs.* time of delivery.

A hidden resource of the information system that was not mentioned in the actual report of the system scientist is the political support of the new system, as well as the "negative" resource of the political opposition. Although the system scientists did not feel it appropriate to assess either the positive or the negative political resources that existed in Sacramento during the study, subsequent history of their recommendation clearly showed the advisability of their having done so. If one fails to assess the political opposition and support, then one may waste many hours of time on carefully contrived plans that meet with a disastrous attack by political conservatives and others who make their political bread out of opposing new ideas.

Here again the system scientist can reply to the criticism by pointing out that the invitation did not include a study of political reaction; such a study is up to the governor's office and other agencies to conduct. But the criticism stands, nonetheless, whoever is to blame. One can scarcely say that a systems approach has been taken if a large part of the design is bound to die on the vine for lack of political fertilizer.

The system scientists spent some time in considering how the components should be designed, because this area seemed to them to be their chief concern. One proposal was to have a large component which would be the centralized library of all relevant state information; all health records, educational records, motor-vehicle records, and so on would be stored in one centralized place. According to this plan, the various agencies and the public could then request information from the central source. The request would be transmitted over lines to the central information depot, which would screen the request, process it through the computer, and send back whatever it was appropriate to return to the requester. In this event, the components of the system would look something like a telephone system with one central office and individual telephones connected to it. The original telephone systems had very much this character before various telephone systems were put together into a larger system. The greater the number of requesters, of course, the more cumbersome the system becomes. Hence the designers of the state information system suggested instead that pools of information be stored in various agencies, and that there be a centralized component containing information about where information was stored. If the user knew how to indicate what he needed by means of some code number or other symbolic device, he could then query the central system, which would tell him where the information he is seeking is to be obtained, or else transmit his request to the appropriate agency. Naturally, in some cases the user could skip the central information system if by common sense or prior knowledge he could easily ascertain where the information was. This means that the components of the system consist of information files in various agencies (perhaps with some duplications) plus a central information component that tells where basic information is stored.

The remaining components consist of the linkages and terminals that connect the information requester to the central information system and thence to the files in which he is interested, as well as computer programmers located at each "data bank."

Evidently the measure of performance of the central information component will be its ability directly to identify for an individual requester where the information he is seeking is stored; the gross measure of performance in this case is the number of correct pieces of information transmitted minus the number of times that the central information system is unable to respond even though the information exists. The data bank at each agency is a component of the system, and its gross measure of performance is the number of times it responds correctly to requests either from the central information system or directly from the user. The gross measure of performance of the linkages and terminals will be their accessibility and the ease and accuracy with which they transmit requests and return the answers.

These gross measures are modified by cost estimates to form net measures of benefit, which are then used as a basis for designing the central processor and the agency files.

From remarks made above it's clear that the designers of this system were not overly concerned about the management of the system itself. They did take on the problem of the generation of a plan as this was what was requested of them. They did not concern themselves, however, with how the plan would be implemented, i.e., with the consideration of the political forces pro and con that might assist in getting the plan into action or might resist it. Nor did the system scientists concern themselves with the overall problems of the management of the new system—that is, they did not set up a management guide that would tell the state how well the system was performing in terms of its realistic objectives, and in terms of the activities of the various components.

Finally, what was perhaps needed most of all was a method of evaluating the new system. I have already indicated that some of the problems of evaluation were neglected because the basic issue as to whether or not information is really needed by the current manual system was largely ignored. For example, it's almost trivial to ask whether the public needs to "inform" the motor-vehicle agency yearly, i.e., whether licenses should be issued annually. Perhaps they should be issued every six months, or perhaps every two, three, or five years. In fact, there is apparently some ambivalence on the part of the motor-vehicle agency on this score, because motor vehicles must be registered yearly, whereas drivers need only register every five years. Note that this problem is the familiar one of balancing opposing objectives, in this case the objective of up-to-date information with the objective of information cost-minimization; for the management scientist, the problem might be solved by a model.

As mentioned in the last chapter, there are a number of very critical evaluation problems besides those directly concerned with the value of the information to the user. There is the question of whether information should always be made available to every requester. As clearly it should not, then what criteria should be used to decide whether the user's request should be satisfied? These are the serious problems of confidentiality associated with state records, especially health, criminal, and educational records.

Although the prospect of large information systems for state, local, as well as federal governments seems bright given that computer technology is obviously advancing quite rapidly, it is clear that in most of the systems approaches to this critical problem many of the basic issues are still ignored. We can become quite fascinated by the rapidity with which a nation's requests can be supplied out of a data bank by a computer processor. Indeed, computers have already become so clever that they can satisfy not only specific requests for raw data, but also can perform various kinds of statistical analysis. They can draw charts, do correlation tests, or tell the user whether or not in certain segments there has been a significant increase in certain types of workers, homeowners, or business firms. All this transmittal and analysis of data is as pleasing as any new technology is apt to be. But just because data can be handled rather rapidly does not mean that a rapid system is a good one. If a system is able to take data and analyze it very rapidly and hand it back to the decision maker when he is in no position adequately to interpret the meaning of the data, then the rapidity will create more harm than it does good.

The proposal for an information system in California was not bad, but it wasn't good either. What surely is needed as a minimum is an information system that will help the policy makers to make their decisions. For example, a few years after the study was completed, a new governor, who was "efficiency minded," was pushing the policy of across-the-board cuts in cost. Was this wise? My guess is that, even had the study's recommendations been fully implemented at this time, the computerized information system would not have helped very much at all in evaluating the efficiency policy of the new governor.

The systems scientists who designed the new system have a good point in saying they had no assigned responsibility for the larger aspects of their system; at best, they might have refused to make a proposal unless they were granted broader responsibilities. Hence the real trouble may lie with the administrators. They seem to expect that system design will be done for them and that they have no active role to play in the design. To the alert management scientist, such a passive attitude on the part of the managers is bound to be a mistake.

The essence of the mistake lies, according to the scientist, in insufficient planning, in creating systems that are too oriented to present needs rather than to needs that are bound to arise in the future. Before we hear the scientist on the subject of values, therefore, we should learn something about his concept of planning. To do this, we need to learn something about decision making over time.

III. Systems Approach to the Future

9. TIME

When we left the management scientist before the illustrative interlude, he was promising to answer his critics by telling us how to make a systems approach to human values. But before he does this, we need to fill in another piece of his story; it is concerned with time, a very important item in any attempt to design human systems.

All along we've been talking about the "larger" system which "embeds" the smaller one, much as a bed embeds a body that itself embeds a heart that embeds a blood vessel and so on. But this language fails to capture the very important idea that the "larger" system may be the future world. In this sense of "larger," the larger system is infinite, stretching endlessly into future generations; it also stretches endlessly into the past, but management scientists are not interested in this sector of the larger system except as a source of data, as they somewhat naïvely think they can do nothing about it.

Now every design we have discussed has a future orientation, but most of the illustrations and models have concerned themselves with the near future rather than the far future. Perhaps a better way to say the same thing is that the designs and models we have considered are all directed towards the next *stage*, rather than a number of succeeding stages. A "stage" of the future is like an episode in a story or an act in a play; the story of Ulysses' encounter with the Cyclops is a "stage" in the epic of the *Odyssey*. Stages are relative to the system, of course. The next "stage" in a production department may be next week's manufacturing schedule, while the next stage in the design of an electricity-generating plant may be 50 years, the life of a generator.

In the example of the port study, the next stage was the duration of the technological improvement of the engineers; in the last chapter, the next stage was the computer-based information system of the state. In both of these cases, the designers could have considered stages beyond the next one, in order to get a perspective of the "larger" system. That they did not do so is evidence of their implicit assumption that the next stage is "separable" from subsequent stages, i.e., that whatever is done next does not hinder or help the changes to be made later on, a somewhat questionable assumption, to say the least.

To look at a system in terms of stages does require that the scientist be reasonably precise about his definition of a stage; but, as we shall see, it is sometimes rather easy to define though not to predict future stages. Often multistage looking is called "dynamic," while single-stage looking is "static." Such nomenclature immediately biases the picture, because in this culture of the red-blooded, courageous, and free, it is always better to be dynamic than static. In point of fact, however, dynamic models may be far less feasible to apply, according to the scientist's notion of feasibility, than static models. Indeed, the management scientist has a very specific philosophy of the future that we need to explain before we look into his method of dynamic, or multistage, modeling.

The first credo of the scientist's philosophy of the future is: *The future is always less certain than the present* (and, presumably, the past as well). Thus each successive stage is vaguer, less predictable, and so it becomes less and less feasible to plan beyond a certain point. Notice, for example, that the years 1984 and 2000, although always present as planning targets, have only recently become real targets for the social planners. I know of no one who seriously suggests we plan for the year 2100, or 3000, or 10,000, though presumably there will be people around then, and presumably how they live (or fail to live) will be partially determined by what we do today. The reason the scientist finds it impossible to go very far into the future is that he believes the error of his measurements increases with time, so that eventually all his estimates become completely unreliable. One should note that this situation does not hold in all the sciences; for example, in astronomy we can predict the relative distance of the planets for millions of years and even estimate when the solar system will freeze over.

There is good reason to doubt this first credo of the management scientist even on his own grounds. The "present" for him is not what he sees directly but what he constructs in his models and his imagination. Since he is interested in redesigning social systems, the relevant "facts" about costs, for example, are all judgments about lost opportunities in the larger system, *including the future.* Apparently, the scientist merely means that he's more at home in the present, that the present feels feasible to plan for.

But he does have a reply to this accusation, in his second credo, which reads: *Benefits and costs both diminish at each successive stage.* The idea behind this belief is the so-called discount factor. Consider planning for the purchase of new equipment, e.g., of automobiles for a passenger fleet. Management scientists use a "replacement model" for such multistorage thinking. Each year (stage), management must decide whether to run the automobile another year or replace it. Past experience provides some good estimates of the likely maintenance costs, inconvenience costs, and replacement costs for each coming year. Now if a cost, like replacement, can be postponed for another year, then the manager can use the unspent funds for other purposes, e.g., he can invest the money and receive, say, a ten-percent return. The same old basic logic of the "opportunity" cost is at work here; the cost of buying something now rather than later is the lost opportunity of using the funds for other purposes in the interim. In the language of the scientist, future costs should be "discounted" back to present values. It follows that, in terms of present values, planned costs for each successive stage will decrease at ten percent a year, if ten percent is what the manager can earn on his investments.

The same logic applies to benefits. If I have to wait ten years for my ship to come home, I forego the use of funds during that period; hence a benefit ten years from now must be "discounted" back to present value.

You can see how this second credo helps to overcome the embarrassment of the first: there is really no point in planning beyond that stage at which the present value of the benefits and costs of the stage are virtually zero. Most equipment-replacement models, for example, run out to ten or 12 years at the most; there is no advantage in exploring beyond that time because the benefits and costs become so small (there are also other technical reasons that the "time horizon" need not be larger).

You may not like this "scientific" philosophy of the future, and with good reason. If replacement models were applied to professors rather than automobiles, a lot of us would be forced to retire earlier than age sixty-five or seventy, particularly since aging professors are expensive and young assistant professors are cheap. The point the aging professors wish to make is that the cost of replacement in terms of present value is always much larger than the monetary estimates would indicate. Perhaps more telling examples are the future generations of society. If we say to them, "Look, a benefit to you must be discounted back to present value," we're saying that the later in history you are born, the less important you are, which is not very nice of us, after all.

Of course, the management scientist is aware of the criticisms, and he points out that his philosophy was meant for the single entrepreneur or decision maker, whereas the retirement of professors or the design of societies of the future involves the values of many people. We promised to postpone the value issue until we had heard more about the scientist's exploration into the future, so let's examine some of his more precise models, after which we can see how he handles some of the vaguer and more complicated realities of the real future.

A rather obvious evolution from single-stage to multistage thinking occurs in waiting-line and inventory problems. For example, most servicing units are tied into other units. When you enter a hospital, your entry sets up a whole series of requests, first at the insurance desk, then at the nurse's station, and so on. A systems approach to hospitals, therefore, should look at the processing of patients in terms of a series of stations; there is no point in becoming highly efficient at one stage if this means piling up a huge waiting line at some subsequent stage. The first stab at a sequencing problem consisting of several service units is to draw a "flow chart" in which the boxes show the service being rendered, and the arrows indicate the routes taken by the customers. If data on arrivals and service times are available, it may be possible to construct a multistage waiting-line model that predicts where bottlenecks will occur and in general helps the manager to optimize across the board. Similarly, in the problem of inventories, one may find a sequence of stages: raw material, in-process, and final. An "optimal" policy at one stage may play havoc with later stages; the scientist's solution is to build multistage inventory models.

In the simplest cases the flow chart will be linear, in the sense that everything or everybody flows from one stage to the next as follows:

But in many realistic situations the flow may be far more complicated and look like the following, for example:

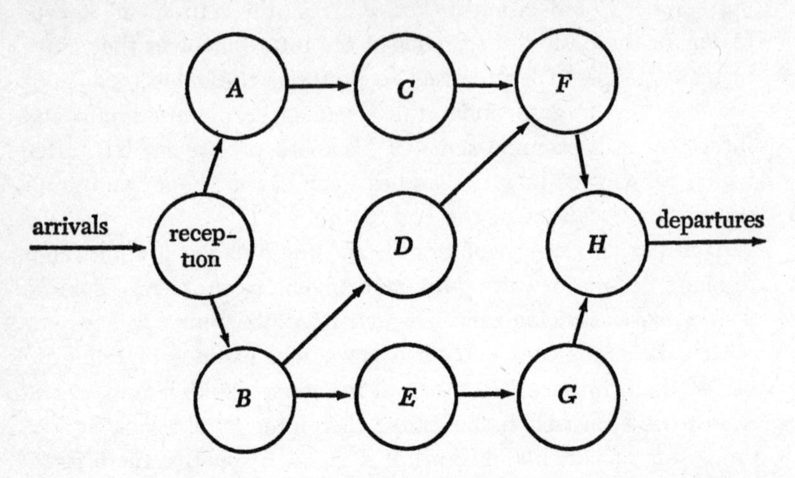

Here at reception the arrivals are divided into two groups, some going to *A* (e.g., severe emergency operations), the rest to *B* (scheduled operations). Those who go through *B* are also divided into two groups and go either to *D* or to *E*, and so on.

These more complicated flow charts are called "networks." If the arrows indicate the flow of people or things, then the manager may be interested in waiting times, idle times, or inventories at each "node" of the network. But he may also be interested in total time or total cost for an item to flow through the system. A very important illustration of this managerial concern is in the area of construction or engineering development. In this case, the nodes represent jobs to be done, the arrows show the sequence, and one can indicate on the arrow the time of the task. Thus the first node in the construction of a house may be the laying of the foundation (there are prior nodes in the whole system, of course, e.g., the drawings, bank loans, etc.) After the foundation is laid, then the house is framed. After this node three jobs may be done simultaneously: the walls, the floor, and the roof. In the flow diagram, this would be shown by three arrows emanating from the same node.

The construction engineer can make a number of important decisions with the help of the network. First of all, he may want to estimate the time to completion. He will find that there are some critical tasks that determine this total time. Consider a very simple flow diagram of the following sort:

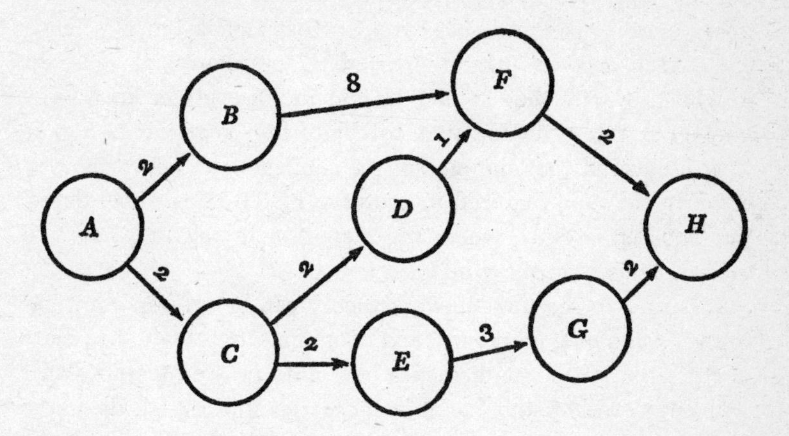

It is easy to see that job *B*, which takes 8 days, is "critical" in the whole task, and hence that the task cannot be completed in less than 12 days. In fact, the "critical" path that determines the minimum time to complete the task is given by the nodes *A-B-F-H*. If it were essential to complete the job sooner, one might make *B* more efficient (at some cost) by cutting its time in half. But then *A-C-E-G-H* becomes the critical path and the job will take 9 days; therefore cutting *B* down below 5 days does not help improve the total time. Consequently, the construction engineer can use the network to estimate the costs and benefits of reducing the time of each job.

Construction engineers and production superintendents have long been familiar with these networks. Sometimes they can solve their problems manually by "Gantt charts," which enable them to shift the sequencing of jobs around in order to approximate a smooth flow. This method is quite limited, however, and has been replaced in recent years by CPM (Critical Path

Method) and PERT (Program-Evaluation-and-Review-Technique), which are based on mathematical models, can be programmed in a computer if necessary, permit consideration of hundreds of nodes, and (in the case of PERT) allow consideration of uncertainties. In fact, PERT and CPM have become so popular that their meaning becomes lost in following their routines, a very common occurrence when management-science techniques become institutionalized. One instance is reported in which, despite the use of PERT charts, a contract kept "slipping," i.e., the estimated time to completion kept increasing. It was discovered that the amount of slippage was approximately equal to the time required to keep the PERT chart up to date—just another instance where the management scientist failed to include himself in the whole system.

It is easy to see how network theory can be an important aid in the design of traffic routes and communication lines. The main contribution of the mathematics here is to provide methods that will cut down on the "combinatorial" complexity. If there are ten ways of designing one node (or route) and ten ways of designing the next node, then there are a hundred ways of designing them both, and soon the number of alternatives gets out of hand. In specific instances, mathematical magic may help to avoid a complete exploration of all possibilities. An excellent example occurs in the job shop, where each day a thousand jobs may come in and be processed at various job centers. The number of possible schedules the foreman could follow is enormous; yet job-shop foremen do run their systems somehow with apparent effectiveness. The fascinating task of the management scientist has been to understand the ingenuity of a good foreman and to see if mathematics can improve it.

Finally, in multistage thinking we should recognize that, even after a plan has been adopted, it may be modified. This comment recalls the idea of cybernetics, a term connoting the steersman. Cybernetics is a mathematical method of evaluating and controlling a process on the basis of its experience. There is an abundant technical literature on cybernetics, especially as it is applied to mechanical systems. The next chapter discusses one of its potentials in social systems.

Thus far, multistage thinking has been restricted to the model or simulation. But if we adhered to this restriction, a great deal of planning would be impossible. As we did in the case of linear-programming models, we can ask whether the *logic* of the model may not be applied to realistic planning situations, even though a significant part of the data as well as a precise model are lacking.

But now we have to admit that we're getting ourselves into some deep waters. Planning—i.e., multistage thinking—is not altogether a popular idea; it connotes "socialism," or the control by the state of our freedoms. It seems to run counter to imagination, genius, insight, creativity, and a host of other "in" words of our culture. The planning enthusiast is a type, and as he thinks he has a hold on the whole system, he's worth listening to before we condemn him.

10. PLANNING

"The way to look at the whole system is in terms of a plan," says the planner, "because the one thing that makes man superior to the rest of the animals is his ability to think ahead—to plan. Planning means laying out a course of action that we can follow that will take us to our desired goals."

For the planner, every person must be a planner to some degree. Every morning it's necessary for me to take my automobile and travel to work. Normally as I start out I think of the route that I will take. There are in fact only a very few alternatives and I usually scan each alternative in terms of the time of day to determine what the traffic load is apt to be and the likely convenience and speed with which I can travel. I then select one and follow the plan. If my planning has been successful and I attain my goal, I arrive at my destination within my time limit and without too much effort on my part.

Of course this simple example is not planning in the sense of the last chapter, because there is only one stage, repeated over and over. But in this example we can discern the essential ingredients of a plan as the planner understands the term. A goal is set, a group of alternatives is created, each alternative is scanned as to whether it will or will not effectively lead to the goal, one of the alternatives is selected, the plan is implemented, and the decision maker checks to see how well the plan worked. The last piece of information will be used to control the operation of the plan, as well as to plan better in the future.

As an illustration of the multistage aspect of planning, look at an industrial manager who is considering building a new plant. Say for the moment that the size and equipment of the plant have been decided, and he must now determine where to locate the plant. He will begin by thinking of what he wants the plant to accomplish in terms of the manufacture and distribution of goods. He will then determine the cost of constructing the plant for each location, the accessibility to personnel, and the distribution costs from the plant. But if he is wise, he will think beyond the first stage of the operation of his plant. He should consider, for example, the possibility that the product manufactured in the plant may become obsolete in ten years; if this may happen, he should design the plant so that the costs of conversion are minimized. Also he should examine the way in which the plant and its personnel may influence the industrial and social development of the region. In other words, the planning manager should try to sweep in as many aspects and future stages as possible.

For planning enthusiasts, it seems almost foolish to ask why one should plan, for they have a whole series of arguments readily at hand in support of their ideas. They point out the absolute need to prepare for all contingencies. The argument is that the decision maker should minimize surprise, because for the planner surprise is an unsatisfactory state of affairs. If I set out in my automobile and really don't think about the route I am taking, I may be surprised to find that certain streets have been torn up and that there are long delays. If I had thought about the matter and possibly inquired of the highway-traffic department, I might have been able to avoid the surprise; this avoidance is a part of my plan. Specifically, the planner emphasizes the need to prepare in terms of money and personnel. The money needed in many large enterprises can be secured only if the details of what one is going to do are laid out carefully in a well-developed plan. To develop personnel it is often essential to train people to perform certain tasks, and this can occur only if we look ahead and determine when and how the training should take place.

In sum, says the planner, the decision maker should be in a position to assess atlernatives before he selects one. Otherwise, if he simply leaps in the dark, the alternative he lands on may not be the one that is most satisfactory, and he will not have had a chance to examine other alternatives in a self-conscious manner. All of which adds up to saying that, if one thinks hard about what one is going to do ahead of time, one is better off.

Despite the fact that planning has its enthusiasts, there is no question that there are also a lot of people who think that planping is either silly or downright dangerous. They point out, for example, that the nations which have gone in for planning in a serious way are the socialist and fascist nations, in which the freedom of the individual has been diminished because of the rigidities of the plan. In these countries people must live in the kind of houses that the planners say they must live in. People must buy the kind of food that the planners decide. Entertainment is all planned. Family living is planned; so is loving, to the extent the planners can influence it by matching mates by computers. Far better, says the anti-planner, to live in a system where freedom of choice determines what happens. To some, the freedom of choice in the marketplace should determine the economics of society. The federal government of the United States should not decide what kinds of light bulbs and textiles shall be manufactured by each consumer manufacturing plant. In general, the federal government should not undertake a large consumer plan in which specific allocations are made for the manufacturing of various items. Instead, the manufacturer should try to develop the best product he can at whatever price he wishes to charge, and the consumers then decide whether they want to purchase it or not. Consumer behavior in the marketplace then determines how the manufacturer will behave in the future, whether he will go bankrupt or be highly successful, and specifically how many items of each product he will manufacture. This is a kind of non-planning activity that freedom lovers take to be so very desirable. It is Adam Smith's "hidden hand" of the free marketplace that should determine the characteristics of the economy, and there is no need of super planners to tell us how to run our country.

Of course, even the economic liberals have to admit that on occasion freedom of the marketplace does lead us into very unsatisfactory pathways, e.g., into unexpected depressions, or into all kinds of illegitimate activities, say in the distribution and sale of drugs. It is for this reason that gradually the federal government has adopted various kinds of regulatory policies which in effect constitute a partial planning of our economy.

Furthermore, the planner is quick to point out that freedom can also be planned, i.e., we can plan those areas of decision making that should be left free from outside constraint: recreation, clothing, homes, and so on. The anti-planner is bound to feel unhappy at such a reply. It looks to him as though the planner wants to be the parent: "It's all right to play anywhere you want, but stay out of Papa's den!"

In any event, as we begin to look at the planning function from the systems approach, we will have to recognize that there is nothing that is obviously beneficial about a planning activity, simply because some group of people will be the planners and we have no special reason to trust them. Indeed, planning is one of the activities of a system, and we must look at it as we've looked at other kinds of activities in this book—in terms of its measure of performance and of its advantages and disadvantages to the organization.

The last point becomes emphasized in the various clichés that surround the concept of planning. For each cliché in favor of the planning idea, there is a countercliché. You should "look before you leap," but of course, "he who hesitates is lost." As one large company urges us, there is no doubt that we should "think ahead," but there is equally no doubt that we should "think now" and on occasion even "think behind."

Some of the clichés, in fact, are utterly empty of meaning. The prescription to beware of rigid planning simply means to beware of the plan that so constrains you that you don't like the plan. The prescription to plan rationally simply means to plan in such a way that you can truthfully appreciate your planning efforts. In both cases, the idea is to plan correctly, not wrongly.

If we are going to consider planning in a systematic manner, we will have to undertake a description of the anatomy of planning as an activity. That is, we must try to break down the system of planning into its component parts. As always, the proper breakdown is no easy matter to determine, but the discussions in the last few chapters suggest a framework which will help to identify the components of planning. Planning is concerned with multistage decision making. Hence it must study (1) a decision maker who (2) chooses among alternative courses of action in order to reach (3) certain first-stage goals, which lead to (4) other-stage objectives. Schematically and in a very simplified form, the framework appears as follows:

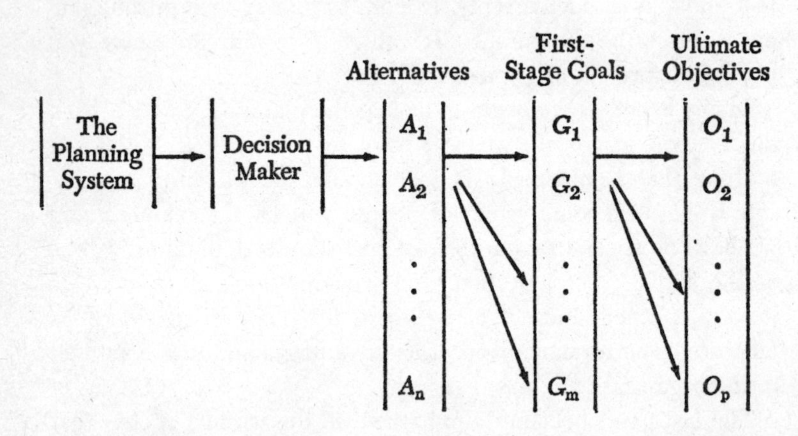

It is the arrows in this drawing that interest us most and help determine the components of the planning system. For example, what does the arrow mean that runs from the planning system to the decision maker? When we think about the matter, we can see that this arrow includes a number of processes: how the planning system fits into the organization, how it observes, how it communicates, and so on. Similarly, as we go across the figure, we will find the need on the part of the planner to describe and predict the nature of the other arrows and entries; these needs result in the following program description of the planning system.

Program 1: Social interaction: the ongoing relationship between the planning system (PS) and the decision maker(s).

Subprograms:

(*a*) Justification (why the PS should exist and its proper role).

(*b*) Organizing subsystem: staffing the PS and establishing its responsibility and authority (i.e., where it "belongs" in the organization).

(*c*) The communication subsystem:

(*c*.1) persuasion ("selling" the plan);

(*c*.2) mutual education (teaching the plan about the larger system, and teaching the larger system about the plan);

(*c*.3) politics (identifying and changing the power structure of the organization).

(*d*) Implementation subsystem (installing the plan).

Program 2: Measurement (identification, classification, prediction, etc.)

(*a*) Identifying the decision makers and "customers" of the larger system.

(*b*) Discovering and inventing the alternatives.

(*c*) Identifying the first-stage goals.

(*d*) Identifying the ultimate objectives.

(*e*) Measuring the effectiveness of each alternative for each first-stage goal (i.e., describing the links between the *A*'s and the *G*'s in the diagram above).

(*f*) Measuring the effectiveness of each first-stage goal for the ultimate objectives (links between the *G*'s and the *O*'s).

(*g*) Estimating the optimal alternative.

Program 3: Test (verifying the plan).

(*a*) Simulation and "parallel" testing.

(*b*) Counterplanning (opposing the plan by its "deadly enemy").

(*c*) Controlling the plan once implemented [see 1(c)].

Of the three major programs of planning, the second (measurement) occupies the most attention at the present time, and this may account for the fact that planning so often fails in its mission.

No one is in a position to give an objective, universally valid estimate of the amount of activity each program should carry on, but nonetheless it may be helpful to suggest some comparative amounts in order to illustrate the systems approach to planning. We shall want to look at the amount of activity both in terms of time (man-hours) and cost, because as we shall see some of the man-hours are apt to be quite expensive if carried on by top executives and administrators. We should also ask the extent to which technology—models, computers, and other applications of science—can assist a specific phase. In the chart below, *H* stands for a relatively high amount, *M* for moderate, *L* for low. Thus an *H* under time and an *L* under cost means that the subprogram ought to occupy a relatively large amount of time, but the cost per man-hour is relatively low. An *M* under technology means that a moderate amount of technology can be applied in the subprogram's activities.

If we were to follow the scientist's systems approach carefully, we would have to describe each subprogram in detail and attempt to assign measures of performance to each. Instead, I shall make some points about each subprogram that seem to require attention no matter where planning occurs and suggest in some cases how one might think about the measure of performance.

Program		*Time*	*Cost*	*Technology*
1(*a*)	Justification	*L*	*H*	*L*
1(*b*)	Staffing and organizing	*M*	*M*	*L*
1(*c*.1)	Persuasion	*M-H*	*M*	*M*
1(*c*.2)	Education	*M-H*	*M*	*M*
1(*c*.3)	Politics	*M-H*	*M*	*L*
1(*d*)	Implementation	*H*	*M*	*M*
2(*a*)	Decision maker	*L*	*M*	*L*
2(*b*)	Alternatives	*H*	*L*	*M*
2(*c*)	Goals	*M*	*H*	*M*
2(*d*)	Objectives	*M*	*H*	*M*
2(*e*)	Effectiveness (alternatives)	*M*	*M*	*H*
2(*f*)	Effectiveness (goals)	*M*	*M*	*L*
2(*g*)	Optimal	*L*	*M*	*H*
3(*a*)	Simulation	*L*	*M*	*H*
3(*b*)	Counterplanning	*M*	*M*	*M*
3(*c*)	Control	*M*	*M*	*M*

The first program, which I dubbed "justification," is primarily concerned with the overall evaluation of the planning effort, and hence one of its outputs will be an estimate of the measure of performance of the planning system. Justification has occupied a great deal of attention in the literature of planning, and especially government and urban planning, where "intervention" of planning may mean a reduction in privacy and freedom. Many of these discussions are couched in moral terms and ask what "right" the planner has to intervene; we shall be examining this question again when we come to the theme of anti-planning. Justification may also mean an economic assessment of the planning function. If the goals and objectives of the organization can be stated in monetary terms, then the economic measure of performance of the PS is the net gain in dollars it produces, "net" meaning total economic benefit minus cost.

The systems question is how much effort should be spent in justifying the planning function, either morally or economically. The generally accepted answer is "very little" (*L*). If a manager thinks he sees the success of planning in other organizations, he will want to have a planning function himself; if he is skeptical or an anti-planner, he won't. If he wants to learn a little more about the problem, there are lots of books and short courses to help him. One thing seems sure: spending a lot of time trying to decide whether to have a planning function is apt to kill the idea entirely. Because it's the top manager who will have to justify the planning function, the cost per unit time is high. Despite a great deal of writing about why planning is good or bad, there is no technology to assist the manager; for example, no one has yet determined the economic benefits resulting from planning.

The next component of the planning function is the one that selects staff and places the function in the organization. What kind of person makes a good planner? Nowadays there are planning curricula in some universities which provide the student with a background in planning techniques, economics, and so on. In addition, everyone recognizes that planners must get along well with people, but not too well, because they need to be forceful at times. Finally, most planning relies heavily on thinking and intuition, and hence these mental functions should be operating well. Beyond these three rather obvious specifications, there is little further to guide in the staffing of planning except prior experience.

This component also has the task of determining how the planning function shall relate to other functions of the organization, and the specific responsibilities and authorities that are given to the planning function. Now, one of the most critical problems of organizing for planning is the potential alienation of the planning function. In many corporations and cities the planning function is carried on in a separate unit. This unit is given the responsibility for developing various kinds of plans, but no authority whatsoever for implementing the plans. But it should be noted that most managers consider themselves to be planners. In the culture of today it would seem quite unreasonable for a manager to declare that he did not look ahead and try to take care of possible surprises that might occur; the public and the stockholders expect him to avoid the unexpected disaster. Consequently, if one inserts a planning function into the organization which carries on the very same kind of activity that the manager thinks is his responsibility, then there is bound to be an alienation between the two functions. Nor can the planners escape this alienation by arguing that they have more time to devote to the problems of the future. The manager is apt to become critical with respect to the planners' suggestions because from his point of view there is an implied criticism of his behavior in what the planners suggest.

The most obvious way to avoid the alienation of the planners is to assign them roles of authority as well as of responsibility. The argument against so organizing is that the planners then lose their objectivity as well as their free time to reflect on the problems of the organization. A compromise is for each manager to assign one of his most responsible people to the planning team. In this design the planning team is still separate from the other parts of the organization but a number of the people in the group have double assignments—both as managers and as planners. Thus the representative of management becomes a kind of hostage; if the managers kill the plan or the whole planning department, the hostage might die as well.

Another solution to the organizing problem is to make the planning function far more obvious by publishing the expenditures that the organization makes in connection with planning, and by assigning these costs to the various divisions and departments of the organization. The planners in effect bribe the managers into accepting their plans, because if a manager fails to accept, he's obviously wasting money. This crude design can be refined by permitting the managers to "sign up" for planning. If a manager does not choose to sign up, then in evaluating his performance the question of his utilization or nonutilization of planning would become an important factor. In this refinement, planning is still doing some bribing of the managers in order to get them to come around to recognizing the importance of the planning function, and there is bound to be alienation.

A final solution to the problem of the organization of planning would be to put the planning function at the very top of the organization, e.g., as a staff function of the board of directors or of the chief executive officer. This has the advantage of permitting the planner to communicate directly with the fundamental policy makers. It has the disadvantage of the planning function's becoming isolated from the actual operations of the organization. As the actual operations of the organization have a great deal to do with how the plan works out, this isolation may prove to be disastrous.

I doubt if any planner today could enunciate a sound principle for organizing for planning or even suggest how an organization should go about learning how to organize. But most of us who have had anything to do with planning feel that it's essential that the managers think about the organizational problem. It's probably a mistake for the board of directors or the top executives to dream up the idea and on Monday morning announce a corporate planning department. But managers have many other responsibilities, and one can't expect them to devote a majority of their time to thinking about how to organize the planning function. Hence the result in the table: a moderate amount of time, the high cost of consultants' and managers' time, and virtually no technology.

What is the measure of performance of the organizing component? A more or less obvious suggestion is that the purpose of organization is to create cooperation in designing and implementing the plan. Suppose we say that the degree of cooperation of *A* with *B* is the extent to which *A*'s activities improve the effectiveness of *B*'s activities relative to *B*'s goals. The organizing component's score of performance is then the degree to which it produces cooperation, a "measure" that can at least be a guide in thinking about the component's activities. To relate this measure to the component's cost, one would have to translate the degree of cooperation into dollars saved by cooperative activity, which is not a very feasible thing to try to do with today's knowledge of how organizations work.

The next three components of the PS are closely related and can be discussed together; their function is to attain the acceptance and if possible the understanding of the plans that the

PS produces by all persons who will have some role in the plan. The three components represent three basic strategies; the first works through persuasion, the second through education, and the third through politics.

The persuasion strategy is appropriate when the planners are convinced that their proposed plan is correct; in such a case, they may want to change public or managerial opinion by using the various tactics of good salesmanship. In such cases, the planners act much as a salesman does when he tries to sell a product to a customer. The tactics of salesmanship have been studied by social psychologists as well as sales organizations, so that there is some technology available here to assist the planners in designing their interviews and briefings. Some of the literature on persuasion emphasizes the point that what is presented to management must appear to be a small increment of change. If the change appears too large, they say, one is not able to sell the product. The idea is that, although the planners may create a much larger structure, what they show to the managers is pieces of the total structure that will move the organization in desired incremental steps. Of course, this kind of "incrementalism" obviously has its defects. It may very well be that the timing is such that a large change is called for. Indeed, it may be that the large change would be accepted, and managers become disappointed in the lack of boldness in the planning function.

More serious, however, is the question of whether persuasion is in fact the most appropriate strategy of the planning function. Many managers rightfully feel that they should understand the nature of the plan and not simply be sold on it; these managers avidly read books about the different role that managers can carry on, about how goals can be set and how they should be related to objectives, and so on. If the organization for planning has resulted in the inclusion of these managers in the planning function, then it would be ridiculous to rely solely on the persuasion process. In such a case, incrementalism might, indeed, be the cure for the wrong disease.

The distinction between the persuasion strategy and the educational strategy can be explained in terms of their respective measures of performance. The measure of performance of persuasion is the degree of *acceptance* it attains, i.e., the degree to which it can influence the public or the managers to act out the roles assigned to them in the plan. The measure of performance of education is the degree of *understanding* it produces, i.e., the degree to which public or managers will adopt appropriate roles because of their own knowledge. Persuasion's contribution becomes very important when (1) the planners have created the correct plan and (2) the public or managers are incapable of understanding the plan, because of time pressures or lack of education, for example. Education's contribution becomes important when the planners themselves need to be educated, and when the public or managers can profitably be educated. It is to be noted that, when the conditions for the need for persuasion do not hold, then a great deal of sales activity in the persuasion component may actually produce a negative gain in the whole system.

In recent years, many planners and management scientists have come to realize the necessity of "teaching the plan" or "teaching the model" about the life of the organization. In situations where the planners must come up with a completed plan before the managers react, it is quite common to see the whole plan rejected. In this event, it is doubtful whether either the planners or the managers learn very much. It would be much as though a wife had to present a plan for the household operations each month, and the husband could reject the plan outright and follow his own wishes.

If management rejects a plan, it does so for certain reasons; these reasons are undoubtedly very important in one's understanding of the organization and unquestionably were not known by the planners, else they would have acted differently. Hence the reasons for rejection are pieces of information that could help educate the planners. The word "symbiosis" neatly describes the healthy relationship between managers and planners; it means that they live together in a mutual-education mode. In this happy state, there is no such thing as *the* plan; rather planning becomes an integral part of every activity of the organization. Planning, as we shall see, tries to measure various aspects of the organization, and as every scientist knows, measurement is an endless process of refinement and prediction.

All of this no doubt sounds rather idealistic in terms of the realities of organizations, because it ignores the politics that make up this reality. The planners cannot expect that everything will be sweetness and light in an atmosphere of mutual understanding, because all planning implies change, and all change is threatening to somebody. Thus, planning always occurs in a power struggle. Any new plan inevitably means to various people in the organization a reallocation of power, just as in the budgeting operation the size of the budget determines the size of departmental empires. Consequently there needs to be some understanding of how people form political coalitions and how a proposed plan will support or weaken these coalitions in terms of the power structure.

Planners are notoriously very poor politicians. Many politicians therefore regard the planning activity to be a waste of time simply because from their point of view the planner is so isolated that what he is proposing to do is politically quite ridiculous. Of course, this phase of the planning is closely tied in with the subprogram of organizing for planning. But politics is something different from the usual notion of organizational structure, because it emphasizes the power relationships between the people in the organization, rather than the explicit lines of authority and responsibility.

Just how this phase of planning can successfully be carried out is difficult to determine. In listing it among the subprograms of planning I am merely trying to emphasize its importance to the planner. He must on the one hand understand his own political tactics, and on the other hand he must be able to discern the kinds of regrouping that are occurring around him. The obvious suggestion is that the planning function contain its own political unit, i.e., a group of people who have some knowledge of the politics of the organization.

It would be nice for planners if there were something called "applied political science," which would tell them how to act in order to overcome political opposition. Unfortunately, no such technology is available at this date; most political scientists either describe public administration or think about politics on the abstract level.

The measure of performance of the political component of planning is the reduction in resistance to the planning function, not through persuasion or education, but through a reconstruction of political alignments. In organizations where political coalitions are very important, the politics of planning needs to be very active: persuasion and education may be hopelessly inadequate to do the job alone. Of course, persuasion does look like politics, but as I have used the terms, persuasion means selling a manager on a specific plan, whereas politics means getting a manager to be "on the same side" or "in the same party" as the planners. In the latter case, the manager and planner agree to cooperate, each supporting the aims of the other.

The discussion has indicated that the amount of activity in the three components—persuasion, education, and politics—will vary depending on the relationship of the PS to the decision makers; the "moderate" to "high" (*M-H*) scoring in the table above is supposed to represent this variation. The cost per unit time is apt to be moderate because many people will be involved in all three activities. Finally, there is some technology for persuasion and education, but virtually none for politics.

The term "implementation" is often used to refer to all three of the components just discussed, but here it is restricted to the activity of setting forth the procedures needed to realize the plan: who should do what, and when. If the plan is large, there will be many steps in its implementation, and these usually must occur in a rational sequence. Consider, for example, all the steps required to implement a rapid-transit plan for a metropolitan area or a computerized information system for a state government.

A great deal has been written about implementation; the PERT technology discussed in an earlier chapter has become quite popular in recent years. I estimate that the implementation activity requires a relatively high amount of time, a moderate cost per unit, and has a moderate amount of technology available, moderate because we still have much to learn about making the implementation smoother. The measure of performance of this component is the time and cost savings of being explicit in the implementing of a plan, rather than relying on people's own judgment and teamwork.

The next group of components fall under the general heading of "measurement." Measurement is sometimes described as the assignment of numbers to things, but it will be far more useful here to define it as the activity of creating precise, accurate, and general information. Precision and accuracy enable us to make refined choices and hence to reduce the risk of error. If I say to you, "Take the bus to get to my home," I am being imprecise, though perhaps accurate because taking some bus is the only feasible way to get there. If I say, "Take the 43 bus at Market and Fillmore leaving at 5:00 P.M. weekdays," I am being precise, but perhaps not accurate if no such bus runs at that time. "General" information is information that can be used in a wide variety of times and places. If the bus schedule changes each day, my precise information may not be general; I could make it general by giving you a day-to-day schedule, so that no matter when you arrived you would know when to catch the bus.

Planning needs to be as precise, accurate, and general as possible in its description of the organization. If its estimates are imprecise, then together all its estimates may be intolerably vague. If they are inaccurate, then the plan will tell managers to do the wrong things. If they are not general, then the plan will only work in a specific context and time, a fatal flaw in many plans.

The diagram of the decision maker, his alternatives, goals, and objectives given above displays the various measurement problems that face the planner.

The first task is to identify the decision maker. As we have seen in other parts of this book, it is by no means obvious who the real decision makers are in the organization, and it is certainly naïve to use labels like "chief executive officer" or "administrator" as *prima facie* evidence that someone is a real decision maker. In this chapter, the term "decision makers" refers to those people who can produce change in the organization; the planner, who is primarily interested in change, needs to know who these people are so that he can influence their choices in one of the ways discussed above. We can see, therefore, how this component which tries to identify the decision makers is closely related to other components of planning, and why a simple identification of "key" personnel as the decision makers may be wrong. In an industrial firm, there may be a number of people besides the managers who can produce change or stop change: members of labor unions, stockholders, customers, secretaries, and so on. In the public domain, the decision makers may include any citizen with enough resources and power to make himself heard.

Many planners, especially those who believe in influencing by persuasion, try to make an "influence map" of the organization which suggests the interrelationship between the people who can produce or prevent change. Sometimes an attempt will be made to assess the degree to which an individual can effect a change, but often this refinement in the measuring process is not feasible. It seems safe to say that there is very little in the way of explicit and safe guides in identifying the decision maker, i.e., the technology is low.

In all the components of planning which are concerned with measurement, there is a common measure of performance. The idea behind these performance measures can be illustrated quite simply by considering again the selection of a route from home to work, which falls in the alternative selection component. Suppose I select route *X* when route *Y* is optimal. Now *Y* is optimal because it contributes the most net benefit to me. Hence the net benefit of *Y* minus the net benefit of *X* is the loss I incur by choosing an erroneous route. The measure of performance of route selection is therefore given by this simple arithmetical difference; if the measure is zero, then the component is performing optimally.

Of course, in most cases we are not able to quantify the net benefits of the measuring components, but the basic idea can still guide us. For example, we can note that in many cases the required measurement is quite *insensitive,* in the sense that missing the true optimal represents very little loss. If good opportunities are very plentiful, or nonexistent, then it may not matter very much what we do. If a plan is patently better, for instance, than what is now being done for all concerned, it may not matter very much who the real decision makers are, because whoever they are they will learn about the plan and implement it. But on the other hand, if top management endorses the plan while middle management sabotages it, then an erroneous identification of top managers as the decision makers may be very serious. In this case, we would say that identifying the right decision makers is very "sensitive." Hence "sensitivity" analysis of all the measurements of planning is required in evaluating the measurement components.

We turn now to the identification of alternative plans that the decision makers can follow. In relatively simple situations it may be possible for a person to determine all the possible ways in which he can pursue his goals and objectives. This usually happens because most of the major decisions have already been made. A manager may have narrowed his choice of a new staff member to two men, so that he knows all the alternatives; but he's already decided to fill the job and what the job should be. Once we turn to the larger problem, the determination of the possible alternatives becomes quite complicated. For example, how does a company know what products it should make? In recent years, many companies that traditionally manufactured food products have started producing a vast number of other products—chemicals for farms, or drugs, or textiles. With this change in policy, the range of possible alternatives is enormous. So, within the planning function there has to be a great deal of thought given as to which of the alternatives are reasonable ones to pursue.

To some extent, logic may be of help here by providing the basis of classifying alternatives. If there is a classification of possibilities, the logician can construct a logical breakdown so that all possibilities are included. The trouble is that the logical classification always ends with the negation of all of the qualities, namely, a possibility that is not *A*, not *B*, not *C*, and so on. But this "negative class" is often the one of chief concern to many managers because it includes new areas in which they might wish to move.

Here again the planner has to leave off being a precise scientist. He needs to encourage radical viewpoints. In fact, I would be tempted to say that whenever planning begins to look as though it is following tried and true procedures that have worked in the past, then planning is in danger of becoming useless. Good planners are continuously asking the most searching, radical, and ridiculous questions (e.g.: Should banks be involved in the handling of cash? Should the post-office department be involved mainly in the transmittal of letters? Shouldn't the soft-drink companies be selling cheap nutrients to foreign countries? and so on). Since there is limited technology available in this area, the best way to proceed is to select planners with radical and unreasonable minds, if you can find them. If not, beware of accepting the planner's version of what you can do and what you cannot do.

One of the difficult problems in creating alternative plans of action is the possibility of a change in the larger system. A redesign of the larger system may make all of the alternatives of the subsystem completely irrelevant. Imagine what might happen, for example, if the engineers working in transportation really could create movable sidewalks and roads. All the careful planning of freeways and automobile traffic within urban communities might then turn out to be largely irrelevant.

Finally, in considering the creation of alternatives, the planners must look very carefully at the research and development function of the organization. Technology can be regarded as a specific way of creating new alternatives. The traveler of 150 years ago might have considered whether to go from Oxford to London by foot or horse or boat, but he also might have asked himself whether or not he could devote some of his energies to inventing a different type of transportation, and thus he might have been on the way to making his fortune in railroads.

As we have already seen, we don't know how to assess the advantages of research, although everyone agrees these days that research output is extremely important in the progress of organizations. Hence, the planners must make judgments about the benefits that are likely to occur as a result of research, even though they lack the data. Planning for research may seem objectionable to those who are in pure research, but the planner believes it to be essential for the survival of many large organizations. The research planning breaks down into planning for basic research, planning for applied research, planning for technological developments, and planning for utilization.

The activity of creating alternatives must involve at least a moderate amount of time, relatively speaking, if for no other reason than that lots of people should get in the act. In most organizations personnel are encouraged to "submit ideas" or even specific proposals for capital expenditures; all these ideas represent possible alternatives, and if the organization really does encourage new ideas, the time spent on creating alternatives may even become quite high.

There is some technology in the creation of alternatives. In fact, we've already seen that, if we can translate alternatives into quantitative terms, then there is the rich technology of mathematics and logic that permits us to lay out the alternatives in a very explicit way. However, there is little technology that guides managers in the selection of relatively new alternatives or in the considerations of the alternatives of the larger system, or in the planning of basic and applied research.

Interestingly enough, the activity of creating alternatives may be relatively cheap per unit of effort, because the activity itself may be carried on by practically all the personnel of the organization. It may be the clerk and not the expensive manager who has the good idea. Sometimes, managers are chiefly concerned with public relations and the setting of goals and objectives and are too busy to spend much time on the creation of new alternatives. Besides, many managers tend to be people who are constantly turning down suggestions, rather than making them.

The next two components of the PS are concerned with setting the goals and objectives of planning. These subprograms are closely related to the program of organizing for planning in that goal setting has to be a function carried on by most of the responsible people in the organization. Because planning is multistage in character, there must be the setting of short-range objectives, which I have called "goals," and the setting of long-range "objectives." Thus the goals are the goals of each stage, in the sense of the last chapter, and a stage can be as short as a month, but is more typically a year. These goals are tied in quite closely to the budget, which determines the resources to be made available in goal pursuit. For the planner the goals must be stated in very specific terms. In the case of an industrial firm they may represent sales goals for each product, cost goals for production, or research and development goals of a very specific technological kind.

The relationship between goal setting and identification of long-range objectives is important. Evidently for most organizations there will be a large proliferation of goals. The purpose of setting the long-range objectives is to put the short-range goals into their proper perspective, i.e., to determine which are the most important goals. As a consequence, the long-range objectives themselves need to be stated in a fairly specific manner, else their relationship to the goals becomes lost and their role as an integrating function becomes meaningless. Thus one of the major aims of the planner in determining long-range objectives is to settle arguments concerning goals in a rational way. For example, one division of a company may urge a sales goal which competes at the budget level with the cost-minimization goal of the production department. The managers must decide which goals should properly be pursued, and the planners hope to aid them by developing an agreement on the long-range objectives. If, however, long-range objectives are stated in quite vague terms ("profit maximization" or "satisfaction of investors"), then the vaguely stated objectives are not a sufficient guide in settling a dispute about goals.

Nevertheless, there may be some real advantage in ambiguity in defining long-range objectives. Long-range objectives constitute the policy of the organization. If this policy is too rigidly defined, then the organization itself may be in no position to modify its basic behavior with a changing environment. As Governor Brown of California put it to his successor: a wise politician keeps as many doors open as possible, and his enemies try to close them. For example, I know of one company whose strict policy was to operate in only one state. However, this rigid and well-defined objective turned out to be far too specific; it made the managers forget that the objective was of their own making and subject to change if they so wished. Many a young executive's inspiration for a market in another state was squelched without deliberation. With the discovery that cigarette smoking apparently is related to lung cancer, the government set out as a long-range objective the education of the public in order to discourage smoking. Having set this objective in concrete, the goals are then all carefully defined. What's forgotten is that the long-range objective, so stated in very specific terms, may be quite inappropriate for human living. The way to solve the problem may not be a direct attack on the smoking habit, but this will be forgotten once the long-range objective becomes so highly specific that people no longer consider whether it should be changed.

The planner tends to go beyond being the scientist at times, when the need to plan becomes greater than the need to be feasible. Consider the range of the plan; how far into the future shall we go? The more cautious of mind, conditioned by the feasible, will insist that we should set our objectives as far as we can see, but not beyond. His analogy might be the wagon train, pursuing its way westward across the country, the leader of the train being able only to plan as far as he could see. For this mind, long-range planning becomes futile when the information that's available can only cover the next few years, and the long-range plan is set for 20 or 50 or 100 years ahead. But the planner objects to this line of reasoning. In the first place the analogy is no good. Evidently the wagonmaster did prepare far ahead of what he could "see," in terms of the equipment that he brought along as well as the weapons that he took to fight the Indians who lay "beyond his horizon." Actually, if one peers into the future, one does so with a great deal of success in certain areas and more obscurity in others. It does not seem unreasonable to believe, for example, that as long as humans are around they will have children. Consequently, 1,000 years from now there will be our descendants. This forecast is a fairly assured one, and if we are doing certain things today that will be harmful to people 1,000 years from now, then we should plan to avoid such policies. To the planner, those people 1,000 years ahead may be more important to us than we are ourselves, simply because they are the natural extensions of our being. Such a thought was certainly behind the thinking of the conservationists at the beginning of this century. These earlier planners thought that the entrepreneurs were depleting the supply of energy from the earth, so that within a few generations there would be no coal or oil left. The conservationists in their own day were planning far beyond what others thought to be the visible horizon. They believed there would be people around a century after them and that those people would be deprived of a very rich source of energy if the existing policies were continued. The same need to plan "far beyond the horizon" occurs in the fighting of disease by various kinds of drugs or the fighting of insects by various kinds of insecticides. The short-run goals become quite clear in both instances, but the long-range objectives may stretch out for centuries ahead in terms of harmful effects on flora and fauna.

Hence, the good planner has to be imaginative and speculative; he has to invent stories ("scenarios") of what might happen in order to keep the distant future alive and significant in the minds of decision makers. Of course, this is also his weakness; if he has something unpleasant to predict, no one will like him.

Normally, as managers and planners converge on a statement of long-range objectives, they find that there are several conflicting objectives. Most companies, for example, recognize profitability as one of their aims, but they also recognize responsibility to investors and public as another, and these may conflict at the financial level. The long-range objective of profitability by itself may imply plowing income back into the business, but such a plan may run counter to the interests of present stockholders who look forward to sizable dividends. Also, most companies today regard themselves to be acting in the public interest by means of the free-enterprise system. The public interest as an objective, however, may come in conflict with some of the specific goals of companies. In America we learned this the hard way; we found that we could not rely in all cases on companies to advertise their goods in the best service of the public, and consequently various kinds of governmental restrictions had to be imposed on the selection of advertising goals by companies.

Whenever the long-range objectives seem in conflict, then the planner must set to work to see how the conflict is to be resolved. Here again, just as ambiguity is sometimes important, it also may be important to keep the conflict running for a long while simply because too quick a decision in favor of consistency may lead to totally unsatisfactory results as far as the operation of the organization is concerned. Again, the wise planner gives up his "scientific" role in favor of political expediency: vagueness and conflict of ideas may become more desirable than precision and consistency.

From this discussion it should be obvious that the chart given at the start of the chapter is a vast oversimplification; the objectives are strung out over many stages of the life of the organization. It is evident that long-range objectives are related to the problems of human values. As we shall see in the next chapter, there is some technology available for "measuring" the values of objectives and specifically for resolving conflicts among the objectives.

In practice the activity of setting goals and objectives is usually conducted in various kinds of conferences between managers and planners. Although goal setting and objective determination require continuous reevaluation, the total amount of time spent may not be great (*M*). But, as in organizing for planning, the activity is bound to be expensive per unit of effort, because it involves the time of expensive people (*H*).

The next measurement activities of the PS try to establish the links between the alternatives, the goals, and the objectives. These activities are essentially engaged in trying to understand the environment of the organization. The planner turns his attention to the following type of question: If such and such an alternative were selected, then what is the likelihood that this or that goal will be attained? This is a problem of understanding the environment because, as our earlier discussion pointed out, the environment, although not controlled by the decision maker, does determine in part how the alternative courses of action are related to the goals and objectives. If a walker takes the wrong path, he may find himself bogged down in a marsh. The environment, in other words, has a good deal to say about the probability that he will have a dry and comfortable walk. In a similar manner, the financial environment of a firm will determine the probabilities that certain goals of the firm can be attained. The attitudes of customers have a good deal to do with the success or failure of sales goals. The policies of nations around the world will determine in part whether the plans of the State Department are successful.

As we have already seen, there are useful technologies for understanding the environment. These were described in the discussion of models of the organization by use of mathematical techniques. Sometimes the computer can be ingeniously programmed to simulate the future of a company or a government agency. In this case the planners can ask the computer to forecast what would happen if a certain alternative were adopted, and by means of various kinds of judgments and facts stored in the computer, a prediction will come out. It is clear that forecasting is an essential part of understanding the environment, because we are looking ahead to determine what the likely fate of a specific course of action will be. There are various kinds of forecasting techniques available, some of them using statistical methods, some the judgment of experts, some a debate between experts. In this last connection, as we shall see, the problem of understanding the environment is closely related to another program of the planning function, namely, the creation of a counterplan.

There is a relatively high degree of technology in this phase of planning. The cost per unit of this phase is slightly above average, simply because the kinds of people who will be engaged in it are rather expensive types of economists, engineers, and the like. The amount of the activity is probably about average, as long as the organization does use technically trained people.

The last subprogram in the measurement phase of the PS is the selection of one of the alternatives, i.e., an estimate of the "optimal" plan. In a way, this might seem to be simply a phase of the last step, the understanding of the environment in order to evaluate alternative actions. It would seem that, if one has correctly evaluated each alternative action, it would be more or less obvious which plan should be selected. But the point of separating this particular program from the one preceding it is to make clear that the actual selection of a plan is based on the evaluation not only of the alternatives but also of the goals and objectives. Just because an alternative is excellent for a given goal does not mean that that alternative should be adopted, because we must consider the relationship of the goals to competing goals. A very subtle point here is that a plan that seems on the face of it to be highly rational may be quite distasteful to a number of people in the organization. Thus, the plan to automate production in an industrial firm may meet severe opposition from the union. This implies, of course, that in setting the goals the planners forgot the need to look at the goal of minimizing strikes, or other types of unfavorable union action. In a way the selection of a plan consists of a review of the steps that have preceded it in order to ascertain how the specific plan under consideration relates to all the goals; often the result of the review is the discovery of a hidden goal. The discovery may then lead to a new alternative, and so on.

If all goes well and hidden goals are revealed, then the selection of a plan from a set of alternatives can be made by technical means. This is so even in the old-fashioned sense of selecting a plan that is cheapest, in which the manager uses various kinds of accounting information and some fairly simple forecasts, and in which he only scans a few alternatives. If there are many alternatives, the problem becomes much more complicated; nevertheless, the existence of mathematical models and computer technology may make this selection of the optimal plan a technologically based activity. The amount of activity in this phase of planning is relatively low, but the expense per unit of effort may be quite high simply because computers and even mathematicians are expensive.

Finally, we come to the subprograms which attempt to test the plan. Clearly, there is bound to be overlapping here, because measurement is a kind of test procedure. But the pattern we have been following is one in which the PS studies the various relevant pieces of the organization, the decision makers, the alternatives, and so on. Finally it arrives at an overall plan, which its piece-wise study implies is the best among the alternatives. Now, the task is to put this final product to a test. The measure of performance of the overall test is the degree to which it improves the accuracies of the measurements of the parts.

I have already mentioned simulation as a technique of measuring the effectiveness of alternatives for goals. In the overall test, its aim is to play out the plan on a computer or in a "gaming" situation in which some of the game players actively try to thwart the plan. With the advent of small but highly efficient computers, simulation and gaming may become an accepted way of testing planning ideas in the future. They should not, however, be regarded as an ultimate test, simply because a simulation is only as good as its designer's intention and knowledge.

Another test of the plan I have called "counterplanning." Very often, this subprogram is not active in the PS, at least in an explicit way. It may be felt that, in the deliberations that are carried on in the other subprograms of planning, enough has been said about opposing plans. The point here, however, is to bring all these oppositions into focus. The need for a counterplan rests on the fact that every plan that appears to be best is subject to many different kinds of error. Some errors will arise from the inadequate setting of goals or insufficient consideration of all of the goals, some because the objectives were not stated clearly enough. Errors will also occur in judging the alternatives available and in understanding the environment. As I've pointed out in the case of management-information systems, the selection of a plan implies a view of what the world is like. Back of every plan that has ever been proposed for a city or nation or industrial firm, there is a story being told about how the world will behave, how the world does behave, and how the world has behaved. The term "story" seems quite appropriate here because, on the one hand, it implies a sequence of events that have been created in someone's mind, and, in the second place, the story cannot be construed as objectively accurate. So many of the judgments that have gone into the setting of goals and their relationship to objectives will never be considered in any ultimately valid sense. So many aspects of the environment are unknown, and so many alternatives will never have been considered. Consequently, the manager needs to see the basic assumptions in any proposed plan. He can do this best if he is faced with an opposing plan of action.

Now a counterplan needs to have certain characteristics. It must appear highly reasonable and attractive. Hence, it must use all the data that were used to build the plan, but must give the data a different interpretation based on its plausible view of what the whole system is like. This is why the counterplan is a "deadly enemy" of the plan: it should make the decision makers stop and realize that some of the arguments for the plan are open to serious question. As the purpose of the counterplan is to prevent errors in making basic assumptions, without this component an organization may build up an unwarranted confidence in a plan simply because so many agree. The value of the component is its savings in avoiding overconfident acceptance.

The activity of the creation of a counterplan does have some technology available in terms of some of the work on "programmed debate" that is being done in social psychology. It is a moderately expensive kind of activity and the amount of it should probably be about average.

Once a plan is implemented, there must be control, which includes feedback of information about the operation of the plan and change of plan when needed. Since there is no such thing as a perfect plan (in the sense of a plan based on objective and reliable information), it is essential that the planning function be designed so that, when a plan is implemented, it is possible to feed back to the managers information about what has occurred and to lay out in general terms the steps that must be taken for change. This is an essential for the proper "educating" of the plan which was discussed above.

As the last chapter indicated, system scientists have noticed the relationship of this stage in planning to feedback mechanisms that occur in the design of hardware systems, e.g., of ships, missiles, and the like. The planner, too, is impressed by cybernetics. He is especially aware of the importance of timing. If one has to wait too long for information about the operation of a plan to get back to the managers, then a course of action may be initiated that is completely disastrous as far as the system is concerned.

Although the technical aspects of cybernetics are not generally adaptable for setting up a design of the feedback and change of plan, nevertheless the concepts are extremely useful. What the planner strives for is something comparable to the cybernetician's "negative feedback," i.e., a situation in which information coming to the manager arrives at the correct time for him to take the appropriate course of action. Positive feedback would be the situation in which information is coming in at the wrong pulse, so to speak.

One can see that this phase of planning does require recapitulation of all the steps, so that, as additional information pours in, correct change can occur. A change of plans is in effect a new plan and must be based on examination of each of the preceding steps, even the step of organizing for planning.

Thus, there is some technology available in the control subprogram of the planning. The amount of activity and the cost vary, depending on how the plan works.

In this chapter we have considered planning from a systems point of view. Another way to look at what we have been doing is planning for planning. You may now wonder whether I've set in motion an endless, cancerous enlargement of the planning function, to plan to plan to plan, and so on. But we can't escape the issue about the total amount of activity that should be allotted to planning, for this is clearly a systems problem. And common sense tells us that in some instances the amount of planning should be large, and in others it should be small. For some kinds of organization, e.g., stabilized institutions like banking, the need for planning may be small. For other organizations in which the activities are based on individual initiative (e.g., research organizations), the amount of planning may also be quite small. But in military and space organizations, in city governments, or in the State Department, the amount should be large.

But these rather commonsensical judgments may be wrong. Perhaps now is the time for banks to start planning in earnest, in order to revamp their antiquated information processing. Perhaps the nation should undertake a systematic planning of research, in order to rationalize the very chaotic and inconsistent way in which the federal government funds research activity.

Perhaps the Department of Defense should reduce its emphasis on program planning. And so on.

From the planners' point of view, these "perhapses" imply that the overall effectiveness of planning must be measured. The obvious measure is the increase in total benefits of the system as the result of the planning activity. Planning must more than pay for itself. It must in fact so pay for itself that the energy devoted to it compensates for the opportunity cost of planning, i.e., for the use of the planning funds in some other program of the system.

The reader may feel that these final remarks introduce a bit of a paradox. The decision as to whether or not there should be planning is itself a plan. The decision about the effectiveness and opportunity costs of planning are again a planning type of decision. But just because there is a paradox need not detract from the value of planning. Indeed, the systems approach itself is based on a paradox. The approach advises us to look at the "whole system," but the amount of effort we spend on trying to understand the whole system is itself a systems problem.

Of course, it is the planner and the systems analyst who land us in this paradox. If you ask the planner why you should plan, he replies, "Pay my salary and I'll find out." Your answer might appropriately be "Forget it." We become trapped in the planning paradox once we get to thinking like a planner. The question we can legitimately raise is whether we *should* think like a planner, and once we raise such a question we need not try to respond by planning thinking.

A less involved response to the planner is to question his specific way of approaching systems problems, and especially the goals and objectives. Is there a planning or systems approach to human values? Let's see.

IV. The Systems Approach and the Human Being

11. VALUES

The practical-minded manager and citizen will long since have asked whether the systems approach has really paid off in practice. Of course, the answer to this question depends on the person who tries to answer it. Many practitioners of management science and operations research cheerfully claim savings in the millions of dollars for either industrial firms or government agencies. Others will frankly admit that, while there have been savings in some instances, in many cases the studies have never been implemented. Others will point out that, although the savings are difficult to pinpoint, nevertheless the whole philosophy of systems approach has introduced a great deal of rationality into organizations so that total effectiveness has obviously improved.

However, any evaluation of the systems approach evidently depends on how we value. Specifically we must turn our attention to what the real objectives of a system are and how the scientist goes about determining them. Unless we know what the *real* objectives are, it is clearly quite impossible to determine whether any approach to the managing of a system constitutes a gain or a loss.

By underlining the word "real" in the sentence above, I mean that, in the matter of stating objectives, people are often deceptive, not necessarily deceptive on purpose but deceptive because they themselves are unaware of what their real objectives are. Stating what we really want is a very personal matter and our statements may have other aims than revealing our real wants and needs: we want to impress people, we want to keep people supporting our projects, and so on. And, naturally, most of the time we don't know what we want.

Consequently, the scientist and planner understand fully that, to "pinpoint" the real objectives of a system, some detailed study is required.

Nevertheless, we should recognize at the outset that not all the scientists and planners feel that it is their responsibility to determine the real objectives of organizations. Instead, the ultimate objectives—the policies of the organization—are said to be the responsibility of the managers. These policies are "given" to the scientist and planner, who then determine the goals of each stage that best serve the manager's ultimate objectives.

The idea that the scientist and planner are not responsible for estimating the ultimate objectives might be called an "engineering philosophy," because in the profession of engineering this idea so often represents a relationship between the client and customer. According to the engineering philosophy, it is up to the customer to specify exactly what is wanted. An easy example occurs when a customer comes to shop at a department store. The managers can reasonably expect that the customer will know what he wants, so that the clerk who is there to serve him can determine whether there are items on the shelf which satisfy the customer's wants. Similarly, if an industrial firm wants to acquire a piece of equipment, e.g., a computer, it must specify what it wishes the equipment to do, and the engineering department or engineering consultant will then try to determine whether such equipment exists or needs to be developed.

Of course, in many cases a customer may not be exactly sure what he wants, simply because he has not been able to make his needs specific enough. Consider that elusive problem of designing a house. It's obvious that the architect cannot resort to the tactic of merely questioning his clients as to what kind of house they want. He must try out various kinds of spatial arrangements and, by getting his clients to react to them, he acquires some deeper knowledge of what their true wants are. The clients also learn a great deal about themselves in this process. Consequently, in architectural design there is a modified engineering philosophy in which the architect and the client try to work out a mutual understanding of the client's real values.

We can see the need to modify the pure engineering philosophy very clearly in the case of the design of computer installations. An organization which is essentially unaware of the potentials of the computer may express its "real" needs very poorly. Hence, the larger computer corporations use "systems engineers" to assist the client in defining his needs so that they fit the reality of the computer more closely.

But even the modified engineering philosophy is not a satisfactory basis for the design of change in organizations. It is not satisfactory because it assumes that eventually the customer or the manager will always be able to make his real needs sufficiently clear so that that scientist and planner can design the desired system, i.e., the system that best serves the objectives of the customer. However, there is the psychological fact that the statement of needs and wants is often confused and frequently wrong, simply because statements of wants and needs serve so many different purposes for the individual. Managers are quite willing to state the positive side of their wants and needs, i.e., the objectives that glow and make their organization appear fine and upright. They want to speak of service to the public, technological advance, dividends for stockholders, numbers of classrooms, amount of throughput of traffic, and so forth. They describe the "objectives" of their organization in terms of these positive values.

But in all determinations of objectives there is the negative side as well: not all the positive objectives can be attained at the maximum level. We have already seen this point frequently occur in the discussion of the systems approach. In the input-output model it is necessary to "constrain" the system in various ways, that is, to impose limitations on various kinds of positive activities. Thus, there will be a limit on the number of students that can be educated or on the salary of the faculty. The industrial manager must admit that he does not wish to exceed certain costs in production and consequently that he is willing to produce items that are defective, or he is willing to have shortages, or he is willing to fire workers, or he is willing to incur strikes, and so on. All these negative aspects of the organizational enterprise have to be brought out in the determination of the real objectives.

But it is clearly quite difficult, if not impossible, for the customer to specify these negative objectives because often he doesn't even think about them, and, if he does think about them, he tries to recast them so that from his point of view they no longer exist. No amount of questioning or probing is apt to reveal the true nature of the negative constraints that the manager is willing to incur.

There are other reasons that mere verbal probing may not show the real objectives. Consider again the case of the design of a house. The client at the outset is not looking at the real system but instead is discussing some ideas about the system as they appear in various drawings. Because he is dwelling in the realm of ideas as opposed to the realm of physical arrangement, what he says as he wanders through the realm of ideas may not at all reflect his wishes in the real physical environment. Indeed, he is often astonished to see how his ideas turn out in the real construction of the building.

Finally, an even more serious difficulty of modified engineering philosophy occurs in those social systems in which there is no opportunity whatsoever to ask the customer what is needed. In the case of the design of a highway system, for example, it is not feasible to ask everybody in the community exactly what his needs are. Of course, the scientists and planners may conduct various kinds of surveys to determine traffic patterns, but these are at best very weak types of evidence concerning the real needs of the citizen. The fact that people take a certain route does not imply that this is the route they wish to take. And even if one asked them what route they wished to take, they would not be able to respond in any effective manner simply because they are unaware of the alternative possibilities. For example, they may not realize the potentials of electric cars as opposed to the present automobile. They may not realize that in the future automobiles may be guided by electronic devices. How could the citizen possibly state his real preferences with respect to transportation when the alternatives have to be presented to him in such futuristic terms? But even more important for most buildings and highways with a survival time of at least 50 years, the real customers are in the future, and, of course, there is no possibility whatsoever of asking them what they need except by assuming that they will be very much like we are.

The proponent of the modified engineering philosophy, therefore, is pushed into the position of saying that he will undertake problems of the design of systems only when the objectives can be stated in a reliable manner. This greatly narrows the opportunity of the scientist and the planner and is a position very hard to justify until we have explored in some depth the possibility of a scientific determination of objectives. If there is a methodology by which the scientist can determine the real objectives of an organization, then it's difficult to see how the scientist and the planner can exclude this phase from their studies. They could only do so if they argued that *properly* the manager must determine the objectives. But how could they understand what is proper without having again looked at the entire system, including its objectives?

Hence, we will assume that the question of the real objectives of an organization is a legitimate question for the scientist and the planner to try to answer provided that they can find some suitable methodology for doing so.

As a beginning, in thinking about the objectives of a system it is natural to ask *whose* objectives are to be served. Since we will be assuming that the answer to this question is in terms of certain people, let's call the set of all such people the "customers" of the system. The customers, in other words, are the people who should properly be served by the operations of the system. In the case of an industrial firm, the "customers" are not only the people who buy the products but also the employees, the stockholders, and perhaps interested sections of the public. In the case of a government agency, the customers are a subset of the citizens.

It's essential that the scientist and the planner identify the customers because only then can they have a basis for determining the real objectives. We note right away that the role we have previously called the "decision maker" may not be the same role as the one we are now calling the "customer." The decision maker is the person who has the ability to change the system, i.e., the responsibility and authority for such change. Evidently the customers of an industrial firm or of a government agency are in no such position. But the scientist and planner will point out that the manager behaves correctly if he serves the customer's needs and incorrectly if he does not. Consequently, it is in some sense the customer who "decides" how the manager should behave. In this sense the customer could be thought of as the decision maker because he provides the base in terms of which the decision making ought to occur in the proper design of a system. In an indirect sense, therefore, the customer is the decision maker, and the scientist and the planner so consider him in this discussion.

The problem of the scientist and the planner is now to determine the *real* objectives of *the* customer. In the simplest case the customer will be a single person who is identifiable and who can be studied in such a way that his real needs will be revealed.

Here we come to a parting of the ways among the scientists. Many management scientists and planners will attempt to relate the real needs of the customer to an economic base and specifically to net dollars accruing to the customer in each stage of the system's operation. Another set of scientists, however, will argue that the dollar alone is not a representation of real objectives and that in addition one must study the behavior of the customer. We'll consider the "behavioral scientist" in the next chapter and turn our attention specifically to the theory of the economic base of objectives in this chapter.

The management scientists who try to relate the real objectives to an economic base have some convincing arguments in their favor. As they point out, we live in an economic culture, that is, a culture dominated by monetary considerations. It's true that the customer does not want dollars just to own dollars, but a large class of his wants can be satisfied by the exchange of dollars. Consequently, the possession of dollars is a sound substitute for these real objectives. In many cases it is quite proper, says this management scientist, to use the dollar as the scale along which the merit of a system is to be measured. Of course, this is not all there is to the story, because we may have to modify the dollar values in various ways, but the modifications will all be in terms of mathematical functions and the basic quantitative unit will still be expressed in dollar terms.

Thus, the economic hypothesis we are now investigating says that the real objectives of most customers of systems are determined by the (modified) net dollar return, and the measure of merit of a specific design of a system will be along a monetary scale.

This idea has already been illustrated in the story of the alcoholism mission of Chapter IV as well as in the planning system of the last chapter. In Chapter IV it was suggested that the seriousness of alcoholism could be measured in terms of lost days of gainful employment, and that these lost days could be translated, in principle at least, into dollar values for the citizens of the state. Note that the scientist and the planner think in terms of net gain rather than gross gain. The net gain is the difference between the total gross dollar gain of the system to the client and the cost of the system for the client. In many cases the scientist and the planner express this idea of net gain in more general terms by talking about the *total benefit* that arises, say, from an Apollo space program or the building of a house *minus* the costs that are produced by such programs. The word "benefit" is used in the economic sense in order to highlight the fact that benefit minus cost can be used not only by industrial firms but by government agencies. It's true that a government agency is not "out to make a profit," but it is out to make benefits for the citizen and the specific benefits that it tries to generate are economic benefits. Consequently, the same basic philosophy of "net profit" that is applicable to industrial firms can be used as well for government agencies, according to the economic hypothesis we are considering.

In the case of industrial firms the benefit cost analysis is an old concept. For centuries accountants have been struggling to represent the values associated with the firm in monetary terms. The operating statement of the firm is supposed to represent the benefits that have accrued to the firm minus the costs. Similarly, many management scientists and economists are struggling to translate the benefits of government services—post office, patent department, health, education and welfare, and so on—into some kind of "national accounting system" with a monetary base.

Cost-benefit analysis as it is currently being practiced in government represents a broader view of the missions of government agencies than was possible under older accounting and budgeting practices. The cost-benefit analyst is interested in determining all the relevant economic benefits that accrue as a result of a given kind of activity. In order to do this he must build at least a crude economic model in which the benefits are represented in economic terms. The purpose of the model is to try to answer the following question: Which costs and benefits are to be included? How are they to be valued? At what interest rate are they to be discounted? What are the relevant constraints?

In effect, cost-benefit analysis generalizes on the management scientist's idea of measuring the performance of a system in dollar terms. To see how such a measure might be created, consider, for example, the design of a new school building in an urban district. The cost-benefit analyst will begin to think in rather broad qualitative terms about the real objectives of such a building—education of grammar-school children, supplying work for teachers and for administrative personnel, upgrading the value of property in the neighborhood, and so on. He then sets to work to try to define these objectives in more precise terms. He might translate the vague objective of "education of grammar-school children" into a specific objective, for example, the completion of the sixth grade at a satisfactory level by at least 100 children per year. In considering economic benefits to the employees of the school, he might decide that the proper objective is to attain a total income produced by the school for teachers, administrators, janitors, etc., of at least 300,000 dollars per year. He might define upgrading of property in terms of an increased average value of real estate in the area of the school of at least two percent.

The management scientist has now succeeded in quantifying the objectives, but the quantities are expressed along different scales. In the first case, the number of children graduating is not yet in the dollar terms used to define the benefit of incomes of the employees of the school. In order to find a common economic unit, the scientist might try to express child education in terms of its economic potential for the community. He might express the completion of grammar-school education in terms of its contribution to the potential income of the working adult. His economic model would include probabilities of dropout, of death, and so on.

If the cost-benefit analyst fails to find an obvious translation into dollar terms of an objective such as the safety of automobiles, he may still attempt to make the translation by other means. Suppose, for example, he could get representatives of the customers to rank the objectives and tentatively to assign weights. If some of the objectives are in dollar terms, then the weights will enable the cost-benefit analyst to translate the other objectives into dollars as well. Thus, if the representatives of the neighborhood rank the objective of educating children up to the level of the sixth grade as twice as important as the potential net income produced by the school, the cost-benefit analyst might feel justified in evaluating the education of the 100 children as twice the net value of the dollar income of the teachers, administrators, etc., in this case 600,000 dollars per year.

To make this example specific, consider some possible objectives of a school system: education, job opportunity, recreation, social meeting ground. A citizen first ranks these objectives and then assigns numbers between 0 and 1 to represent their relative values for him: for instance, education (1.0), job opportunity (.5), recreation (.1), social (.05). In making these judgments, says the scientist, the citizen is performing much as he would if asked to judge the physical weights of various objects; he is saying that for him education is twice as important as job opportunity, and the latter is five times as important as recreation, and so on. In this exercise, the management scientist is taking a step in the direction of the behavioral scientist of the next chapter by attempting to examine how people react in various situations. When he asks people to set weights on the objectives, he is in effect asking them to express the sacrifices they would be willing to incur. Thus, when someone says one objective is twice as important as another, he is saying that he would be willing to give up or trade "two units" of one objective for one unit of the other.

It is perhaps a surprising but nonetheless true fact that people seem quite capable of making such judgments, both about physical and value "weights." Whether the latter are as reliable as the former often turn out to be is a matter of debate. But we should note that, if there is some reasonable confidence in the customer's value judgments, then the management scientist can usually convert all objectives into economic terms.

Both the cautious scientist and the indignant humanist will

sense real difficulties here—the scientist, because the judgments need to be more carefully specified; the humanist, because it all looks like a trick of "scientism." The cautions of the scientist can be heeded, in practice, to yield more reliable economic translations of objectives; whether the outrage of the humanist can also be taken care of remains to be seen.

As a further check on his economic estimates, the scientist may examine past decisions on the part of the managers. He recognizes that in any rational past decision, the manager himself must have made implicit judgments about the relative values of his objectives, and as some objectives are bound to be economic, the scientist can also evaluate the others in economic terms. For example, most of us recognize that it is very difficult to determine the dollar loss of a personal accident, but if the scientist can determine how much money is put into safety devices on automobiles, aircraft, roadways, and the like, he may be able to infer from past behavior what is the implicit dollar value of a life or a limb that has been assumed by the managers.

So far we have been assuming that the scientist can identify a *single* decision maker. But in all of the illustrations we have been talking about many customers rather than one. Only in the case of the department-store purchaser was the single customer apparent. In this case it looks as though the manager had every right to assume that the person to be served by the system at that point in time was the single individual who arrived at the counter. Clearly, in the case of the design of buildings or schools or hospitals, there are a large number of customers, and it would seem very unsafe to say that there exists a single person who truly represents all of these customers' interests. Or, if the scientist were to make such a statement, he would have to determine the correct manner in which customer representation is to be made. Consequently, to say, for example, that the school board is *the* customer of the educational system is not to solve the scientist's problem in any way. The scientist operating under the systems approach must convince himself that the school board is properly representing the true customers, and in order to do this he must examine the real values of many customers.

We can begin to see how the complexities of reality have created serious problems for the would-be management scientist and planner. Things went very well when there was but a single recognizable decision maker and a single stage of decision making. Call this an SS (single stage, single decision maker) problem. Matters become more complicated, as we saw in the last chapter, when multiple stages appear, but still there was in that example a single decision maker (MS problems). But here we are confronted with a multiple decision maker and multiple stages (MM problems). It is no wonder that many scientists and planners struggle to find a *representative* of the multiple decision maker. The obvious method is to find one customer and expect him to fight the politics of MM later on. This escape, however, is a snare and a delusion, because the single decision maker cannot stand for the multiple customer unless the scientist can justify that the single customer represents the multiple customer. Sometimes, as we shall see, the scientist does not look for a real customer, but an ideal or abstract customer concocted out of the multitude of conflicting interests. Even so, how does he justify this construct of his mind?

The scientist will reply that to focus on the single decision maker is a way to get started. If at the outset he tries to focus on the morass of multiple decision makers, he will never be able to make even the first approximations to the solutions of the problems. Call this starting point a myth, if you like. For example, it was even a myth to say that the customer who walks into the department store is a single decision maker, because most customers represent many different kinds of pressures within the family. Furthermore, this single person standing at a counter is himself a complex of minds, conscious and unconscious—id-ego-superego-feeling-sensation-intuition-thinking—minds with unique internal politics of their own in each self.

Of course, the idea that all real problems are MM problems is no news in the story telling of this book. As soon as the program planner tried to find data on the alcohol mission, he found that the "simple" problem of collecting data is itself an MM problem. In general, the problem of how to design a systems approach is always MM in all of its phases. Still more generally, the design of any inquiring system is MM. That's why the meaning of "science" is still so obscure despite all the wise talk about it. (More on this point later.)

But the recognition that all real problems are MM need not stop the scientist so long as he can take all of the relevant interests and combine them into one unifying objective, i.e., one unifying decision maker. As I have said, he will admit at the outset that he may be wrong.

Actually, the problem that the scientist faces does not seem to be different in kind from the typical problem of any science, because all science exists in a state of uncertainty about many aspects of the world. Science itself must proceed from approximation to approximation. The scientist does not expect to be correct because such an aspiration is not feasible. Instead, he uses the method of "explicit assumption making." In the application of this method to the study of systems, the scientist or planner attempts to the best of his ability to determine a single decision maker in terms of a multitude of customers and their interests. In making his judgment he also makes explicit all the assumptions that he has made so that he himself, as he proceeds in the systems design, may continuously reconsider his assumptions and so that other scientists or planners may react to his assumption making in an effort to improve it. For example, the last chapter suggested some strong methods of testing the assumption of the planner.

The process that we have just described is very similar to the method used by the experimental scientist. Any experimental scientist is aware of the multitude of factors associated with his experiment that he cannot possibly control. What he does not know at the outset is whether these factors will have some influence on his experimental results. If they do, they may so confound the data as to make any kind of inference impossible.

What does the scientist do in such a situation? What he does is to follow the same "explicit assumption making" procedure described above. He sets down as clearly as possible what he assumes to be the correct state of affairs with respect to the uncontrolled variables. As the experiment proceeds, he can test whether the assumptions are correct; furthermore, other scientists who examine his results will know exactly what he has assumed and will be able to test his assumptions. The idea is that science progresses by continuous modifications of its basic assumptions. At each subsequent stage of science the assumption making is improved. Science will never reach the ultimate ideal of the correct answer but, by the method of explicit assumption making, it does learn more and more accurately about nature.

To some extent at least the method of explicit assumption making has worked out reasonably well in the physical sciences. Can we expect a similar success in the design of social systems? There are some serious reasons that we might not expect to find any progress occurring. These all rest on the question of what a real improvement in the method of designing systems is supposed to be like. More generally, how does one determine whether there has been an improvement in the understanding of social systems? The problem of the multiple decision maker is to determine how the multitude of decision makers can be unified into a "representative" decision maker. In the language of welfare economics, the problem is to take the various interests ("utilities") of the human individuals who are the customers of a system, to translate these interests into quantitative terms, and then finally to create a single measure which represents the unified social preference. The process is very much like voting for political candidates. Presumably there are many different opinions and wants of the citizens. Each citizen goes to the poll and votes for a candidate who most nearly represents his desires. The candidate who gets the majority of the votes then stands for the unified decision maker.

Of course, such voting is a crude expression of social values, because it wipes out intensities of need or desire, every voter being counted exactly as one. Furthermore, in most democratic societies it is easy to criticize the manner in which the citizen is informed as well as the manner in which alternative choices are made available to him. In more detailed and deeper efforts to design social systems, the scientist or planner needs to try to unify the variation among consumer interests in a much more specific and rational manner. But then the question arises of how he shall weight the interests of various customers. Is it appropriate in the first place to compare the wishes of one individual with the wishes of another? In other words, even if one makes an explicit assumption that comparison of values is legitimate, how can one possibly justify this assumption under questioning?

Economists for decades have been examining the problem of comparing utilities, chiefly because in Western culture there was the expectation that it might be possible to generate a basic economic measure across society in which the wishes and needs of each citizen could be adequately represented in quantitative terms. The critics of this position have argued that it's impossible for the scientist to make such comparisons of individual values because he has no way, so to speak, of "getting inside" the heads or hearts of individuals in order to make the appropriate comparisons. The critics have argued that the only way people can express their wishes is either by verbal statements or by certain types of behavior, and that in neither case can the scientist succeed in making direct comparisons.

In recent years there have been some rather clever attempts to overcome these critical objections by use of probabilities. The basic notion is that, if one can observe the risks an individual is willing to undergo in order to gain an objective, then one can make a comparison on the basis of risk aversion or risk taking. This effort, of course, if carried on in terms of behavior, would take us into considerations of the next chapter and far beyond the typical economic approach to problems of value. Indeed, from the point of view of the behavioral scientist, the notion that one cannot compare utilities is certainly naïve, and a behavioral scientist will argue that there are many instances in which it's quite feasible to compare individual values. Managers, he points out, do this kind of thing all of the time, when a city government decides, for example, to build a school in one district rather than in another. Furthermore, from the point of view of psychology, the intercomparison of values goes on inside the same mind because psychologists recognize that every mind is made up of conflicting minds each with its own value system. The values of the unconscious mind are certainly not the same as the values of the conscious mind, and yet somehow the total mind makes a comparison in order to arrive at a unified expression of its wants.

But even though it may be legitimate to compare human values, it's not clear at all how this comparison will result in a unified representative of the multitude of decision makers. The problem becomes even more critical when we consider the future of the subsystems. Because most important systems survive beyond the generation that creates them, the customers of these systems must include the people who are not now alive and therefore not capable of expressing what their wishes will be. It's interesting also to point out that the customers of systems are the people of the past. Our grandfathers have a definite interest in the kind of world we live in today and in the world we are trying to create. Their voice is with us even though their bodies may not be. Consequently, the unified representative must include all of the interests of past and future individuals who cannot be "tested" directly by the planner or scientist.

The future, as well as the past, represents one strong argument against the idea of a marketplace of decision making or a political arena of choice in which each person expresses his values either in terms of the money he is willing to sacrifice or the political power he is willing to express in his vote. The existing consumers and citizens can scarcely be regarded as the representatives of either the past or the future customers of systems. In the free marketplace these will be the unheard voices, unrepresented and unheeded.

Indeed, very much the same story could be told about the

single individual in the act of making his present choices. Is he truly representing his past self as well as his future self? And yet, it is these past and future selves which really constitute the kind of person he is, even from an economic point of view.

Thus, the scientist's method of explicit assumption making doesn't seem to work well in the context in which the meaning of the social system stretches into the past and into the future. How will a contemporary scientist be able to check the assumptions of another planner or scientist?

The management scientist's answer must be that, however difficult the task, it's essential that we arrive at a coherent and acceptable version of what the future of human systems will be like, in terms of technological innovations, national and international politics, the economic development of nations, and so on. This interest in the assessment of the future has become quite popular in recent years. Indeed, in both France and England there has arisen a movement to examine the nature of future societies, and in the United States a commission was appointed to consider the state of the nation in the year 2000. All of this future-looking activity is an attempt to answer the critics of contemporary science and planning who argue that the interest in technological innovation and the curing of today's problems as such may very well lead to a worsening of conditions in future generations.

The question, however, as I have said, is one of methodology. How can future wants and future conditions be properly estimated? One notion is that there are some intellectuals who have spent a great deal of time considering the nature of society and that these men therefore are in the best position to express an "expert" judgment concerning future conditions. In order to bring out their expertise in a clear fashion, it may be necessary to structure their deliberations in some specific manner, e.g., by letting them interact with each other's judgments and reformulating their judgment in a sequential manner. This the purpose of Olaf Helmer's "Delphi technique" (see Suggested Readings at the end of the book) for forecasting by means of expert judgment. The Delphi technique might be amplified by putting the experts in an atmosphere of debate so that their implicit assumptions can be made explicit.

The critic nevertheless will argue that the entire effort may be a misguided one, simply because what is being examined in the future is the wrong kind of thing. His feeling will be that the determination of economic benefits is only one aspect of the total value picture. If the management scientist replies that it's his responsibility to supply the economic aspect of the situation and the managers' to "fill in" the other relevant aspects, then the critic has every right to claim that the separation is a spurious one. It may not be possible to look at economic benefits in isolation from other kinds of deeper human values of recreation, of safety, of family relationships, of friendships, and so on. Furthermore, the management scientist should explain how the manager is able to make these noneconomic value judgments.

What guides the management scientist and planner in their thinking about systems is always the feasible. The economic values are much more feasible to look at than are the more elusive "humanities" values. The economic values seem to be "out there" in explicit form, either in dollars or in the more tangible aspects of technology, like equipment and service. The hidden, human values are "in there" and cannot be adequately ascertained in such a manner that one can use them in the redesign of social systems.

But this adherence to the feasible is exactly the point that the humanist will wish to attack. The feasible and the explicit may not be the correct basis for human decision making. Those who typically try to approach reality through the spectacles of the feasible are those who create the ugly monsters of our current technology. They are those who forget the really critical human conditions of poverty, distress, mental illness, and the like. They sweep out in one large housecleaning all of the real aspects of human values that are so dusty from their point of view that they cannot be figured into anything except waste. The most arrogant of the feasible-minded actually believe that, by considering economic values, they can eventually handle all of the intangibles in an implicit manner, much as though the human being was an economic being and that all of his values are tied into his basic economic interests.

Interestingly enough, the debate at this point has become a debate about values, namely, the values of those who wish to change human systems by means of science or planning or some other intellectual method. The values of the individuals who view the design of systems from the economic point of view are the values of the explicit, the precise, the "rational." They believe themselves to be most allied with the practical-minded managers who like to see a problem laid out in an explicit form so that they can understand it. The more elusive, intangible parts of values, therefore, are to be ignored or else handled by personal judgment. The values of their opponents, the humanists, are the deeper, individual human values arising from the real feelings of each person. These cannot be represented, says the humanist, by any kind of explicit assumption-making method. They cannot, because the real values of a person are not the kind that can be determined by any kind of "investigation," by either scientist, planner, or anyone else.

But there is another kind of scientist who tries to bridge between the economic-feasible approach to the change of systems and the humanist demand for the representation of "real" human values. This is the "behavioral scientist," a man dedicated to investigating what the human being is like in terms of his behavior. This scientist is less interested in model building than he is in the empirical determination of what human beings do and how they make up their minds. It is his belief that the empirical investigation of human behavior will eventually lead to a sound understanding of the nature of the human being and his societies. Once the soundly based empirical findings have been accumulated, then, say some behavioral scientists, we will be in a position to plan more adequately for human development.

It is essentially a difference between reason and sensation, between the rationalist and the empiricist. It is, as I have said, a difference in values, and in order to understand the values of the empiricist and how they may eventually tie into the systems approach, it is necessary to give a brief survey of the findings of behavioral science.

12. BEHAVIOR

The behavioral scientist is an individual who believes that, by observing how people behave in their environment, it will be possible to describe their minds, ambitions, and frustrations, and eventually to see how these all fit together in one large pattern. Once the empirical investigation has reached a successful end, then the nature of human societies will be understood; consequently the basis for the design of human societies will be the hard-core data about human behavior, rather than the assumptions of models.

To speak of "observing people in their environment" sounds much like the language of a biologist who carefully observes the behavior of living forms. And the behavioral scientist finds much inspiration in the history of biology. After all, it was only in terms of very detailed collection of data about living beings that evolutionary theory was able to grow. Biology did not develop from an *a priori* notion that living forms should display themselves in a hierarchy of living beings. Instead, careful observation was essential before even the idea of evolutionary theory could be created, according to the behavioral scientist.

Perhaps one way of describing the behavioral scientist's approach to systems is to say that he has really inverted the management scientist's or planner's approach. The management scientist sees the nature of the whole system as a determinant of individual behavior. For the behavioral scientist, on the other hand, the "whole system" is made up of the behaviors of the individual persons. Once individual and social behavior have been examined in detail, then one can discover in the observations of behavior the nature of the whole human system.

Because the behavioral scientist is so intently interested in observing how people behave, he is correspondingly less interested at the outset in the difference between the good and the bad, or between the efficient and the inefficient. He wishes explicitly to keep the problem of evaluation out of his observing system lest his own evaluations distort the information that he receives. He wants to be the disinterested observer, a role that he firmly believes has been well established in the physical sciences. The experimenter in the physical laboratory does not regard himself to be part of that which is being observed. "No more," says the behavioral scientist, "should I be a part of the social system that I am observing."

At the very outset, therefore, the role of the behavioral scientist may seem unsatisfactory to the humanist. The behavioral scientist may in fact begin to look something like a snoop. Indeed, says the humanist, the best way to be a disinterested observer is to use hidden tape recorders or to wiretap each home in the community. Thereby the most "objective" data about human behavior will be collected.

The problem of the disinterested observer will plague the behavioral scientist throughout our discussion of him. In fact, in some sense the problem makes his role ambivalent. On the one hand, he wishes to study human behavior, but on the other, he realizes that in order to do so, at times he may have to become deeply involved in the lives of those whom he observes.

So how shall we study human behavior? In order to keep the motivation of disinterestedness at a maximum, the most likely methodology would seem to be that of laboratory experimentation. A chemist brings a piece of material into the laboratory and does various things to it, the material reacts in various ways, and from these reactions the chemist produces some basic information about the nature of the material. By analogy, therefore, the behavioral scientist should bring people into the laboratory, where he can carefully control the variables, do certain things to the people, and observe how they react.

The laboratory has become a common tool of many behavioral scientists. People are asked to do fantastic kinds of things within the laboratory setting. In early experimental psychology, they simply lifted weights, looked at lights, and made judgments of intensities of sensation. Nowadays, the human subject is asked to solve problems, to react with other human beings, and even on occasion to undergo electric shock.

There is one group of behavioral scientists who are very much interested in the nature of human conflict, because it is such an important ingredient of large social systems. The subjects of their laboratory experiments are therefore put in a conflict situation, e.g., one in which they play a game in which one subject must be the winner and the other the loser. In these experiments, the behavioral scientist wants to test some of the hypotheses of "game theory." Game theory purports to provide rational strategies for human beings who must behave—as in games—according to prescribed rules. The rules state the choices of each player and the "payoffs" that occur as a result of each choice.

A very simple game is one in which each player has a choice between two moves, *A* or *B*. In such a game, the payoffs might be as follows:

AA (Both choose *A*):	5 cents paid to player no. 1, 5 cents lost by player no. 2
AB (First player *A*, second *B*):	zero to both players
BA (First player *B*, second *A*):	zero to both players
BB (Both choose *B*):	5 cents lost by player no. 1, 5 cents paid to player no. 2

Common sense says that the first player will avoid choice *B* like the plague as he either gets nothing or loses; similarly, the second player will avoid choice *A*. Hence the "rational" choice of the two will be *AB*, in which neither wins anything. In a laboratory setting with real players, one would expect to find this commonsense result repeated.

The game described above is called "zero sum" because the total payoff to both players is always zero, no matter what choice they make. The more interesting (and perhaps more realistic) games are non-zero sum. One example of these has fascinated the behavioral scientist. It arises from a story told about two captured criminals, Merrill and Anatol. The sheriff tells Merrill that, if he squeals on Anatol, he (Merrill) can get off with a light sentence, while Anatol will go to jail for life. The sadistic sheriff whispers the same deal to Anatol: "Squeal and you'll get a light sentence, while Merrill will get life." If neither squeals, they both get medium sentences. If both squeal they both get life. This "prisoner's dilemma" can be captured in a simplified version by the following game (same notation as above):

AA: Both get one cent
AB: *A* gets 5 cents, *B* loses 5 cents
BA: *A* gets 5 cents, *B* gets 5 cents
BB: Both lose 3 cents

Note that there is no "commonsense" choice, but that *BB* seems all around to be the fairest solution. But now if the first subject can convince the second to play choice *B*, the first may then "defect" and switch to *A*, thereby producing *AB*, in which the first gets 5 cents and the second loses 5. The empirical question, therefore, is which is valued more, the economic gain to one of the players or the removal of conflict. By observing how people react in the laboratory to a prisoner's-dilemma game, the behavioral scientist believes that he can determine the relative values of conflict and cooperation for human beings. In this way, so goes his reasoning, he may be able to supplement the systems approach based purely on economic considerations by more realistic considerations of the values of cooperation and conflict for decision makers. Digging deeper along the same line, the behavioral scientist discovers that the preference for cooperation over personal economic gain, or mere conflict for its own sake, depends on other psychological characteristics; thus there are "conflict types" and "cooperative types," the well-known "hawks" and "doves" of the current political scene. This turn in the investigation goes under the label of "psychological correlates," the idea being that there is no underlying consistent pattern of behavior throughout the human species. Instead, we humans can be broken down into types, and the behavior we exhibit under various laboratory situations depends on our type.

The critics of such laboratory experiments are quick to point out the unreality of the laboratory setting. They claim that the subjects in the laboratory are not necessarily responding as they would normally do in the outside world, but rather are responding to the laboratory environment and especially to the experimenter. Some of the subjects may be highly cooperative, in which case they try to do what the experimenter wishes them to do, albeit at times unconsciously. Others are highly uncooperative and try to ruin the experiment. In either event, what is being "discovered" is not typical kinds of human behavior but rather the reactions of individuals to a highly controlled, constrained environment.

It is because of such criticisms that many behavioral scientists have shunned the laboratory and turned instead to real organizations. There they try to describe what "really" goes on in organizations in terms of a detailed "case history." The investigator identifies an important problem of the organization and studies how people react to the problem in terms of both personal relations and the politics of the organization. He writes out these reactions in detail, and the story becomes the basis of discussion. The discussion itself may eventually lead to certain "principles" of organizations, e.g., that the span of control of the manager should be limited to fewer than seven people, that it is essential to provide a basis for "group motivation," and so on.

The laboratory scientist often criticizes the case-history method rather violently. He points out that in each case history an enormous number of variables have been neglected simply because the case historian could not possibly have had a knowledge of all the critical aspects and may in fact be telling entirely the wrong story. The case historian is very much like any other historian. He must separate out from the welter of historical events those particular ones which he thinks are important. He has no notion whether these really were the important events of the time, and consequently the story he tells may be thoroughly distorted, deceptive, or just plain wrong. Therefore, says the laboratory behavioral scientist, it is essential to study people under controlled conditions, at least to check the accuracy of the case-history method. Of course, the astute case-method researcher may point out that the laboratory investigator can also be fooled; what he thinks is happening among his subjects may not be the really important events.

This is not the end of the debate, however. There are those who feel that the correct way to look at human behavior is to take the single individual and measure various aspects of him in isolation from others. This can be done without having to undergo very rigid laboratory control situations, e.g., in the "free" atmosphere of an individual's home or office, by means of both questionnaires and open-ended interviews. Thereby, says this behavioral scientist, one can determine the basic attitudes and opinions of individuals and eventually go on to the study of individual preferences, a subject of much importance in our discussions in earlier chapters. Thus, in our culture we have seen a large proliferation of public-opinion testing, determination of personal attitudes toward the church, education, conservation, and so on. Some scientists think we are in a far better position today than were our ancestors to understand the basic political attitudes of the citizen. They make very fine distinctions about the far left and the moderate left, the middle, and so on, based on what the behavioral scientists believe to be a sound empirical study of the attitudes and beliefs of individual citizens.

Of great importance in the improvement of the systems is the measurement of human values by means of observing human behavior. Note now the important difference in attitude between the economic scientist and the behavioral scientist. Whenever a value like safety, recreation, or education seems to transcend economic considerations, the economist of the last chapter will struggle to reduce the "intangibles" to economic scales by one of the methods we discussed there. The behavioral scientist, on the other hand, wishes to start from scratch, without making any presuppositions about the value of money or any other commodity. He wants to see how people behave when placed in an environment of choice. When they make a choice, this is taken as behavioral evidence of their values. They may, in fact, forego money for some other kind of commodity, and this choice can be determined only by observing their behavior. Nevertheless, the empiricist at this stage does have to make some assumptions. One basic assumption he makes is that the alternative choices that a person can make in a given environment can be ranked. There will be a "top choice," which is preferred over all the rest, and a "bottom," choice which is the least preferred, and in between in rank order will be all other choices. Of course, the same assumption is made by the economist; in his case, he assumes that more money is always more valuable than less money.

As I mentioned in the last chapter, some rather clever uses of risk taking and risk aversion can be used to translate these rankings into what is called a "utility scale." In the use of the utility scale, the economist and behavioral scientist are much alike, the only difference being the economist's desire to ground the scale in economic value. But behavioral scientists, being freer in their conceptualizations, concoct all sorts of other value scales, consumer preferences, attitudes, "basic values," even the "value" of a way of life. These scales are supposed to describe the true values of an individual person. In the application of this empirical method to the management of systems, the managers would be presented with various objectives of the system. In order for any sensible judgment to be made, the objectives would have to be "pinpointed," i.e., made very explicit. For example, in the design of an urban community the objectives might be spelled out in terms of the amount of recreational space and facilities, the number of students graduated from grammar and high school, the amount of police protection, the amount of "throughput" of traffic on the streets, etc. The managers would then be asked to rank these objectives and, by various kinds of questionnaire techniques, the behavioral scientist would hope to assign values to the objectives that accurately represent the manager's interests. This empirical approach to the identification of the objectives of the managers could then be used as a basis for developing the measure of performance so critically required by the management scientist.

Many behavioral scientists today would hesitate to apply their science in this manner, feeling safer to work in less confusing environments than management decision making. But in addition to the technical difficulties of such an application, there is the question of whether a *stated* preference *per se* means very much in terms of real values. Furthermore, the preferences must be made over an explicit set of objectives, so that many of the hidden objectives are not represented. As was mentioned in the last chapter, the unheard voices of the past and the future certainly do not get into this empirical setting. The behavioral scientist will reply that the stated preferences must be regarded as only one kind of evidence of value. The evidence must be supplemented by a number of other findings, e.g., the actual choices a person makes in "real life" situations. At this point the entire methodology becomes quite fuzzy. After all, statements are "behavioral choices," so that if someone *says* he prefers *A* to *B*, he is making a choice, perhaps as significant a choice as physically taking *A* instead of *B*. Unless we know a great deal about a person, we cannot tell by mere observation of his behavior just what his choices mean relative to his true values.

Perhaps of all the developments in the area of behavioral values the one that is most relevant to planning is the concept of a level of aspiration. The idea is that, although in principle there may be an optimum solution to systems problems, the human being seeks only a certain level beyond which he does not want to go even though there might be economic benefits that exceed the costs. A good example is the person who is looking for a house to buy, but does not try to explore all the possible houses available on the market. Instead, he sets his level of aspiration at a certain point, and then, if he finds a house that sufficiently pleases him, he buys it, even though he might realize that there are better houses available.

Consequently it has been suggested to the management scientist that, instead of looking for the "optimum" solution to systems problems, he consider solutions in the more realistic terms of the behavioral scientist. The behavioral scientist argues that the "correct" design of a system should be geared to the level of aspiration of the customers of the system and not to the idealistic optimum in the economist's sense. A good deal of debate has occurred in the discussion between the two scientists—the economist and the behaviorist—about levels of aspiration. From the economist's point of view it looks as though the behavioral scientist is merely stating the economic fact that it costs something to search among alternatives. If this cost of search is included, then the whole concept of the level of aspiration is adequately represented in the economic models. It would be absurd for a manager to go on searching for a solution when the additional refinements do not pay for the expense of search. The economist, therefore, claims that the behavioral scientist has introduced nothing new in his empirical findings that was not already included in the economic considerations. The behavioral scientist, on the other hand, replies that the level of aspiration is an integral part of human behavior and it cannot be translated into economic terms.

From the systems point of view, the debate between the economist and behavioral scienitst fits into the larger consideration of how the manager ought to spend his time—what issues he should pay attention to and in what depth. Every manager comes to realize that he cannot pay attention to every "important" matter; few managers are self-conscious enough to understand why they pay attention to some matters and not to others. The economist would say that this problem fits into his general scheme of the allocation of scarce resources—in this case, the manager's time—and is solvable in principle by an allocation model. The behavioral scientist would argue that the problem depends on the basic psychological characteristics of the manager—his level of aspiration for various tasks. If both economist and behavioral scientist are explicit enough, they might resolve their differences by means of a general model of managerial time allocation. But, of course, there are many managers who would regard such an effort to be largely irrelevant.

So far we have been discussing studies of human behavior either in the laboratory or in interviews and questionnaires. We now turn to those studies which are essentially examinations of social groups and their behavior in their "natural setting." An intermediary type of study is called "gaming" (not to be confused with "game theory" discussed above). In gaming, the managers are placed in an environment that somewhat simulates their natural one, an idea not unlike that of animals in a modern zoo. Thus, industrial managers make decisions in simulated business games, and ambassadors "play out" the weighty problems of their politics in simulated international games. The behavioral scientist sees that in such games managers acquire an increased ability to see how the whole system works and to abstract the more critical elements of their decision from the less critical (for example, international negotiators in the simulation can discuss the essential aspects of international problems without having to worry about their own country's internal politics). The critics of gaming argue that the seeming reality of the game may be thoroughly deceptive. Their point is that the human animal can be observed reliably only in his natural setting. Here again is the debate between the hard-core empiricist and the more theoretical scientist. It is, of course, a debate about the system of science itself.

Among the studies of humans in their natural setting, perhaps one of the most relevant from the point of view of systems are those which describe sociometric relationships, i.e., how people get along together in groups. The social psychologist is curious as to why it is that certain groups seem to succeed so well in the formulation of their ideas, whereas others never get anywhere. One suggestion is that in the successful groups one or two people assume a strong leadership which keeps things going in the right direction. The opposing point of view is that the successful groups develop their own kind of sharing of ideas and do better in relatively unstructured situations. The designing of groups that work well together has been a major objective of the social psychologist. The contribution of these studies to the systems approach rests on the fact that all system designs have to be produced by group task forces of one kind or another. Consequently, says the social psychologist, it is tremendously important to understand how individuals in a group work together.

Furthermore, when the scientist or planner has developed his solution to a systems problem, he has to interact with the manager. This is a problem that has been mentioned earlier in discussing the organization for planning. The suggestion made there was that the managers play an active role in the planning organization. But from the point of view of the social psychologist, this is not enough. One has to explain how this role is to be created. It is important that the manager feel the recommendation to be a result of his sharing in the activity of the whole group. If he does feel this, says the behavioral scientist, then implementation of the solution is apt to occur. If not, then the alienation between the planner and the manager is apt to be so great that no implementation will occur.

Few management scientists can deny the great importance of implementation. Although there is no data on the number of successful implementations that have occurred in the last decade, there does seem to be a great deal of evidence that many studies have simply died on the vine with a consequent waste of many man-hours and dollars as well as huge disappointment on the part of the people who have put so much of their psychic energy into the effort. The failure to implement recommendations, says the behavioral scientist, arises out of lack of understanding of the human being on the part of the management scientist, oriented as he is to economic considerations. The management scientist, says the behavioral scientist, often fails to understand that there are basic psychological resistances to change, that, if change is suggested by someone external to the organization, then there will be a natural resistance, much as the body tries to resist the implant of a new organ. No matter how excellent the new organ may be, the bodily chemistry is such that the organism sets up a reaction to it. In the same way, no matter how excellent the recommendation may be of the management scientist, people in the organization will react negatively to his suggestion unless the management scientist takes steps to create an atmosphere in which alienation is diminished.

In fact, some social psychologists argue for what they call a "sociotechnical system," a system that recognizes both the psychology of the individual and the technical aspects of the system. This effort is representative of a number of attempts to try to bring the technological and the social sides into a coherent package.

In addition to these social-psychological studies of group relationships, there are the larger studies of cultural bias and the role of language in the understanding of systems. The idea is that into every human society certain basic attitudes are built that cannot be changed without some total kind of revolutionary or evolutionary development. It has been pointed out, for example, that the way in which we talk about nature obviously influences the way we understand nature. The application of this linguistic principle to the study of systems implies that in Western culture we have a typical way of talking about our systems and that our manner of talking about them therefore influences our way of understanding them. We have seen very much this kind of thing happening in the last chapter, where the approach to the objectives of a system was framed within economic language. To a person in the Western world, this is a very natural way to talk about the system, i.e., its economic objectives and the resources by which it attempts to attain these objectives. But the common language of economics that we all use may very well influence our insight into the nature of systems. Whether the behavioral scientist would be correct in inferring that with a quite different language one would have a quite different systems approach is, of course, a matter that is up to debate.

Beyond the cultural anthropologist and the linguist lie the even broader studies of politics and law, fields which are virtually unknown by the management scientist and the economist. The theorist of law would undoubtedly claim that the lawyer does grasp the system, is able to analyze it, and does give the client proper advice about his behavior within the system. Probably most lawyers would agree that the formal court procedures of law play only a very small part in the process of law. An adept lawyer knows the nature of the legal system and can recommend courses of action for his client accordingly, but just how he does this is difficult to transmit to the nonlawyer. Nevertheless, in recent years some behavioral scientists have become quite interested in studying various procedures within the law in order to see whether it would be possible to represent the way in which the lawyer interprets human behavior.

We can begin to see from this very brief excursion the wide spectrum of interest of the behavioral scientist, starting in depth with the nature of the individual person and broadening out into social groups, society, and cultures. The behavioral scientist studies an individual in terms of the choices he makes, the goals he seeks, his beliefs, his concepts of reality, both conscious and unconscious. He sees the human being as a social individual, studies the nature of society and its behavior, and sometimes he dreams of carrying on his empirical studies to a determination of ultimate human values. What is it that all human beings basically wish to have? In the old days, men said that the purpose of a system is to create happiness for the humans within it. But for the behavioral scientist "happiness" is a term devoid of meaning. Furthermore, it's not even true that people "seek happiness" or even the "greatest good of the greatest number."

Nevertheless, it may be possible by extended study of human individuals and societies to find those very basic things which all humans want, say, technological advance, education, knowledge of the world, cooperation, and perhaps beyond this some noncooperative values, an urge for change or for destruction or even for evil. Thus, on top of the cake of the behavioral scientists is the fluffy icing created by the speculators who wish to go beyond empirical findings and make vast leaps ahead to infer what it is that specifically characterizes the fundamental needs and ultimate purposes of the human being. These speculators are akin to the grand cosmologists of the physical sciences, who guess about the origin and ultimate destiny of the universe. Most behavioral scientists and management scientists are sceptical about all of this speculation, though they recognize that, in their own work in systems, they, too, have to make assumptions about ultimate ethical values: after all, even the sceptic must accept an ethical dogma to the effect that the aspiration to know is wrong.

Off and on I have been speaking of science as a system; in connection with the discussion of this chapter it is interesting to point out that the science which studies systems is itself not a very integrated system. In fact, the behavioral scientist who, as I have been saying, is intensely interested in the nature of social systems, rarely speaks to the management scientist, and vice versa.

Why is it that the behavioral scientists are not well integrated with the management scientists? One would think that the two complement each other. The behavioral scientist provides a rich empirical base and the management scientist provides the structure that can employ this base for making inferences about the changes in social systems.

Probably the best answer is that there is a basic psychological difference between the intellectual who goes into management science and the intellectual who goes into behavioral science.

It is the difference between the rational type and the empirical type, or between rationalism and empiricism. The rational type finds the ultimate foundation of his work in the model, in a theoretical structure that shows how pieces of reality are put together in a precise, coherent fashion. The empirical type, on the other hand, finds his ultimate reality in what he directly observes going on around him. The rationalist is aghast at the immense amount of data that the empiricist seems perfectly cheerful about collecting. To him the empirical behavioral scientist goes out with his tape recorder and other devices, comes back with an enormous amount of information, and only then begins to worry and scratch his head about how to analyze the information. From the rationalist's point of view, this kind of thinking should have taken place at the very outset. How does the empiricist know that his data are any good unless he has already made up his mind what is critical and what is not critical in the system? The empiricist, on the other hand, sees in model building an abstract activity bearing no relevance to the real-life situation. He feels that he is much closer to the manager, and indeed in many cases he is. The empiricist can talk directly to the manager in his own terms, become friendly with him, and feel that the manager understands what he is up to, whereas the model builder is apart because the manager himself is not a model builder and doesn't understand what model building is all about.

In recent years there have been some attempts to develop a bridge between the economic approach to systems and the behavioral approach. The difficulty with the economic approach, as I said in the last chapter, is that it leaves out so much that is really relevant with respect to human values and systems, aesthetics, recreation, health, and so on. Its attempts to translate aesthetics, recreation, and health into economic terms seem to leave out the realities of each of these human values. The suggestion has been to develop "social accounting," an explicit technique of evaluating aspects of society from the rich background of the behavioral scientist while keeping the precision objective of the economist in mind.

It's difficult to know whether social accounting will succeed in providing the bridge, but even if it does we have some reason to suspect that the coalition thus created still may not represent an adequate approach to systems. It may not because the coalition of behavioral scientist and economist still approaches the system from one fundamental viewpoint, a viewpoint that might be called the "planning philosophy." It is the viewpoint that by the use of reason and observation it is possible to lay out the structure of a system and decide what changes should occur that best serve the customers of a system. Is the planning philosophy an appropriate philosophy for human systems? If you say no, then you are someone who believes in anti-planning. To the rationalist, it's hard to see how anyone could ever accept an anti-planning philosophy. Indeed, what could an anti-planning approach to systems possibly mean? Well, let's see.

13. ANTI-PLANNING

In some sense all of the approaches to systems that I have discussed thus far in this book are not really commonsense approaches to systems change, or would not have been considered commonsense approaches several decades ago. The idea of using behavioral scientists and planners to assist in analyzing systems and changing them (although as I said earlier it does have a historical background) has not been a particularly popular idea in the United States. The popular idea of how to approach a system is to get somebody to manage it. The manager is supposed to be a person with rich experience in the system and with a perceptive, brilliant mind. The manager examines various aspects of the system, receives some data and reports from the staff, and then makes up his own mind what should be done. This is certainly an "anti-planning" approach to systems as I have defined planning earlier in the book. The manager in most cases cannot make explicit what steps he has taken and he feels no need to do so. The idea is that he can be judged in terms of his performance; if a young man indicates signs of being perceptive and a good leader, then he is promoted. If not, he never climbs the ladder. In this anti-planning practical school, education takes place within the system and is never made explicit.

Anyone who has had experience with managers in American industry should easily recognize this anti-planning idea. In each industry, the manager who has grown up as a railroader, steelman, lumberman, automobile man, "knows the business" and cannot see how some green outsider could tell him anything significant. Such a manager would never think of asking a scientist how he should spend his time or what he should pay attention to; these are matters for experience and "sound judgment" to decide.

Of course, the management scientist and behavioral scientist both feel that there is a widespread myth about "excellent" managers. Naturally, certain people have attained eminence in society for various reasons, but if one analyzes in depth the process by which they made decisions, it is hard to justify that they are great and perceptive managers. Even the so-called great presidents of the United States are a matter of personal opinion. The popularity of Lincoln and Washington may arise not as a result of their own ingenuity, but from the creation of a public myth. It may very well be that Chester Arthur, the least well known of all American presidents, could be regarded as the greatest "manager," as it was under his administration that the United States government was transformed from a politically dominated system into a civil-service system. In other words, the management scientist would argue that the greatness of a manager can be determined only after one has studied the system by building a model of it and comparing what the manager did with the optimal.

So there is one kind of anti-planning approach and its opposite, namely, the practical approach of experience coupled with intuition, leadership, and brilliance *vs.* the analytical approach of the scientist.

Perhaps two more devastating anti-planning concepts are those so often expressed by the sceptic and the determinist. The sceptic firmly believes that we can never understand even minor aspects of a system. He therefore believes that everything that we say about systems is largely a myth invented in order to carry on various kinds of conversations and entertainments. As the nature of the real world is a mystery, he says, we deceive ourselves when we think we are improving anything. It's true we go around shuffling things from one point to another, but in the end, if you try to evaluate whether there has been any beneficial change as a result of this shuffling, the sceptic believes that it is impossible to do so. He laughs at the absurdity of those who think that, for example, transportation is "better" now than it was in the days of our forefathers. He points out that a freight car moved faster in the days of the horse and cart than it probably does now on modern railroads. But even if freight moved faster, so what? Has fast movement really benefited the human being? Can we show that any technology has really proved beneficial? What is the evidence that the technological "spinoff" from science has resulted in more benefit than cost (detriment)? It was marvelous indeed to discover drugs that reduce pain and save lives, but look what harm drugs produce in the human race. Today we move faster, dress faster, eat faster, recreate faster, kill faster than any animal on earth has ever done before. To the sceptic the enthusiasms of the technologist appear to be just one more manifestation of the silliness of a human being. The sceptic is usually the ultimate pessimist.

Of course, the sceptic is an arrogant fellow. The easiest thing in the world is to be a relativist, somebody who says, "It all depends" and "We can never know the ultimate answers." This is something that every student who has ever dug deeply in a social problem will say. It is the mark of the sophomore in the intellectual enterprise. The one thing the sceptic rarely does is to defend his own scepticism. He simply shows the extreme difficulties of answering questions, and as a consequence he regards the difficulty as evidence of his own sceptical philosophy. To the serious-minded, this kind of relativism serves little purpose and is socially irresponsible. Not that this attitude on the part of the serious-minded will in the least deter the sceptic. His approach to systems is that there is no sound approach, and that's that.

A more serious opponent of planning is the determinist, the man who believes that major human decisions are not in the hands of human decision makers but in uncontrolled sociological forces. Recall that earlier we went in search of *the* decision maker and discovered that he was hard to find; sometimes he was many people—all the citizens that have been, are, and will be. But for the determinist, there is no decision maker in the sense of a person or group with an ability to choose: no one or many ever set the policies of an organization or country. For example, the growth of science for the determinist is determined by military and industrial trends in our country, trends that themselves are the products of other social forces. He sees the advent of the New Deal, the New Frontier, and the Great Society as manifestations of underlying discontent, and the attempt of the Democratic party to answer discontent is itself determined by basic political forces that are not in the control of any person or group of persons.

The determinist is not a new creation of our society. Ever since the time of the Greeks, people have been arguing that the events in nature are fully determined and are outside the control of the human being. If the world is fundamentally deterministic, then of course, it would be foolish to claim that by planning or in general by any kind of thinking we can do anything as far as the change of systems is concerned. The changes are brought about by forces outside of our control, no matter how convinced we are that we "make decisions" by our own free will. Planning is simply a guessing game in the determinist philosophy.

The management scientist's reply to the determinist is to try to educate him about the scientific theory of evidence. Sophisticated management scientists will agree that, if a manager really thinks he is a decision maker because he sits in his office and signs a paper, he is probably being naïve, because there is no real evidence for this belief. But the management scientist does believe that, if he conducts a study in depth, certain individuals will appear more likely than others to be the ones who produce the changes that occur; on the basis of enough evidence, the management scientist feels justified in calling those "decision makers" who have a choice. The point is that, in all science, one deals not with final answers but with estimates; hence the assignment of the label "decision maker" to a group of people is an estimate, to be modified in the light of further evidence. The determinist, says the scientist, has taken the obvious error of our estimates and converted it into a basis for rejecting our methodology.

Just as there will probably be no final answer to the argument between the rationalist and the empiricist, there will be no final answer to the argument between determinists and nondeterminists. The determinists often regard the whole attempt to study systems by means of science as a natural product of a highly militarized and industrialized society. In other words, the determinist will subsume the management scientist under his own theory. The management scientist, on the other hand, will regard the determinist as someone who has arrived at his position because of psychological disappointment.

There are two anti-planning positions, however, that need to be considered quite carefully. These are not strictly anti-planning in the sense of the positions just outlined, but find their base for understanding the entire system in something other than the economic and behavioral approaches described earlier in this book. The one is the religious view of the world and the other is the view of the world as a reflection of the self. I have labeled them "anti-planning" simply to emphasize that they both would argue that the scientist's approach to systems is misguided.

The religious approach says that the real planning of the world lies in a power or mind that is greater than the mind of all men combined. It is a world (or cosmic) force. In the case of optimistic religions, it is a world force working for good. Once this notion of a god has been introduced into the realities of the system, then one's attitude toward the whole system must change. It is no longer up to a human being to try to decide on his own what the basic values of each person are and thereby to develop a rational approach. Rather the human being must learn what the god's plan is and try to adjust his behavior to it.

Those who believe in the religious point of view have a very strong argument against the pure management scientist and the behavioral scientist. Recall that I have been saying throughout the book that, because the management scientist and the planner cannot possibly believe that they have the correct plan, they must keep thinking of their activity as a series of approximations in which each approximation in principle is better than its predecessor. But why should such a series of approximations lead anywhere? What is the guarantee that, struggling as we are in the dark, we will find our way into the light? The guarantee, says the religious world view, is some kind of superior mind that assures us children of darkness that the pathway does exist, and he who helps himself will be helped by his god.

Now the tradition of Western science has been one in which the existence and properties of God are not of concern to the scientist *as a scientist.* There are various reasons that the scientist arrived at this viewpoint of the system of science. One reason lies in the fierce political fights that broke out when science announced its intention of divorcing itself from religious doctrine in the fourteenth, fifteenth, and sixteenth centuries. Another arises from what is called the "positivistic" attitude of scientists, namely, the notion that the scientist believes what he sees, and as he cannot see a god, he cannot find empirical evidence for his existence.

Nevertheless, the point is well taken that the management scientist implicitly assumes a guarantor of his activities if he sincerely thinks that he is doing something to improve systems, i.e., if he is something other than a sceptic or a determinist. Indeed, for many management scientists the religious problem of the guarantor or a superior mind begins on reflection to look very much like the very problem that he himself is facing. It may in fact be a matter of nomenclature. For the person who was brought up in religious thinking, the nomenclature is "God"; for the person brought up in scientific thinking, the nomenclature is "progress" or "approximation." The method of explicit assumption making discussed in an earlier chapter does have its implicit "god," i.e., the full expectation that an estimate made under incorrect assumptions will be corrected by other scientific workers, and that this process will continue by increasing our knowledge of nature. In fact, what makes the management scientist something other than a sceptic or a determinist is his belief that the world will remain safe for scientific progress.

Of course, there are many scientists who never consider the social and political foundations of science; they want to work on problems that interest them and have no concern *as scientists* for the future of society or the environment that will sustain human learning. These "pure" scientists are among the strongest of the anti-planners: pure research must not be planned. But if pressed in debate, these scientists would have to admit that the future of society is an important matter for pure research; they assume that someone else, in a nonscientific role, will create the safe society. This assumption, for the advocate of the religious approach to systems, is simply an act of faith on the part of the pure scientist.

Thus, to the philosopher of religion there can be no question that the approach to the whole system on the part of most inhabitants of the world is through a religious world view. Consequently, for him it would be foolish indeed if the planners and management scientists ignored religious world views by concentrating too hard on the economic view of the world. To date there has been no real confrontation between the religious and the economic, partly because both sides wish to keep their own independence, partly because there has been really no need yet for a conversation. One can expect, however, that, as the system scientists become more conspicuous and begin to make some very explicit assertions as to how the world is to be run, they will come into conflict with various religious positions. In some sense they have already come in conflict with those religions which believe that the matters of human decision making should be left largely to the individual and should not be planned by society.

The management scientist, with his firm foundation in an advanced technology, may believe that most religious world views are antiquated. The story of the Aztec culture well illustrates the scientist's point of view. In this culture there was the firm belief that the affairs of men were guided by the gods. Too, the gods themselves were not unified into one supreme decision maker, but rather had their own conflicts. Therefore, in the Aztec culture the approach to the system consisted in trying to appease the gods by various kinds of religious rites, and especially human sacrifice. But in appeasing one god another god might become irritated, with consequent harm to the human system. Hence, the Aztec rulers tried to understand their system by understanding the system of the gods. Along came a highly "advanced" society, the Spanish conquistadores with their technological instruments, and they put an end to the quaint religious views of the decision makers of the Aztec culture. In somewhat the same manner, the management scientist may think the advanced technologies we are creating today will put an end to any quaint ideas about the manner in which a god or gods influence the systems in which we live.

This scepticism about the traditional religious outlook is all the more reason for a confrontation between religion and science in the context of the design of social change. There clearly is no more reason for religion passively to adapt to technological change than there is for science to adapt to traditional religion. The critical issue for the systems approach is to identify the religious assumptions implicit in any proposal for change, be it the poverty program, a war, or cost reduction. When such a confrontation of religion and science takes place, the meaning of planning will change, and the religious type may cease to be an anti-planner; he may be an anti-planner today only because planners do not include the determination of religious assumptions in their list of planning activities.

The second anti-planning approach is based on the analysis of self. It is the position that the world as it really exists, exists in the individual self. As a consequence the planning of total systems is nonsense unless by this term is meant the fullest expression of the individual self. To those who take this point of view, which is the point of view of the inner life, the problem of living consists in the attempt to understand what we are really like in ourselves and the different kinds of selves that we are. There is the power-dominated self, the master who wishes to overcome and, in the process of attempting to overcome, becomes himself the slave of his own domination. Or there is the conservative self, who wishes to keep the world as it is, to keep his possessions and to keep his ideas. And he is overcome by the revolutionary self, in whom revolution is generated out of the very activities of the conservative. There is the annihilated self, for whom all existence becomes trivia. There is the immediate self, who finds its value in the here and now, and completely denies the meanings of ends and means. Or there is the visionary self, who looks for the savior but often finds the devil. In all of this searching for the self there is nothing that looks in the least like the speculations of the management scientist and the behavioral scientist. Indeed, for the self-seeker it is totally inappropriate for the behavioral scientist to classify people into types by observing their behavior, as may seem implied by the phrases "power-dominated," "conservative," "revolutionary," etc. What a person *seems* to do in the eyes of another is actually in the self of the observer, not in the self of the observed. In other words, the "results" of behavioral science tell us mainly what behavioral scientists are like, not what people are like in general. The "recommendations" of the management scientist are an expression of his inner being and have nothing to do with "optimal" changes in reality.

So runs the philosophy of the self. In this philosophy there is no talk of systems, components, and measures of performance, or indeed improvement in the scientist's sense of the word. Whatever improvement is going on in the pictures of the self is an improvement of the person's understanding of himself and has nothing whatsoever to do with the notions of progress that are implicit in management science and planning.

To the scientist the challenge is either old hat or meaningless. Science has long since learned the need to disentangle the personality of the observer and model builder from the rest of reality. Every observation, no matter how carefully made, has something of the observer and something of his instruments within it. The problem, in the scientist's language, is to separate out the "invariances," those characteristics of observation which remain no matter who is observing with what instrument. Such invariances, says the scientist, are not attributable to the self. Sometimes, however, they are attributable to a "group self," when the experts deceive themselves because they all agree. Just as it has long since been recognized in science that we must get rid of the "personal equation," so now the scientist recognizes that we must get rid of the "social equation." But once we do, what remains invariant is our best estimate of reality. If the self philosopher still claims that reality is "beyond" these invariances, he must be talking nonsense. The scientist has struggled long and hard to make his concepts "operationally" precise, and he is not about to regress by acknowledging the meaningfulness of self in any but the purely operational, empirical sense.

To this rebuttal, the self philosopher will respond that the physical sciences may have discovered the invariances, but the social sciences have not. In particular, he will say, the attempts to remove the personal and group equations from the so-called measure of human value have all failed. He finds that every aspect of the economist's and the behavioral scientist's approaches to values are no more than a reflection of their own selves and have nothing whatever to do with the reality of human values in general. Consider, for example, that sacred "axiom" of both the economist and the behavioral scientist which declares once and for all that human values can be ordered from highest to lowest. Why should this be so? True, our backward culture has forced upon us the need for trading *A* for *B* and hence has made us "order" our value system. But in the real self of many people, this niggardly way of expressing deep human feelings is far from reality. The scientist will reply that we could never plan rationally if we gave up the ordering of values, because what would "optimal" possibly mean? To this the philosopher of the self responds, "So much the worse for planning; you have reduced it to its obvious absurdity."

If the confrontation of science and religion has been weak, the confrontation of science and the self has been weaker still. Only in psychoanalysis does the debate flourish, but very few think of applying psychoanalytic theory to the "systems approach." It might help a great deal, for example, if "poverty" could be defined in something other than economic terms; we might then discover how many poor people there are in our rich culture.

Finally among the anti-planners there is the completely nonintellectual approach, the approach that does not believe that thinking in any of its senses is important in the development of human life. It is the approach that finds the essence of value in the song, the painting, the vision, the myth, the feminine, and ultimately the unspoken. What is not said at all is the most important thing of all. Since the management scientist, the planner, and the behavioral scientist spend all their time speaking, then it must be the case that what they spend their time on is the least important part of human life.

Here the confrontation is deepest of all. What shall we say to the person who thinks that speaking distorts human feelings? Shall the scientist say that his point of view represents the absurd, or that it represents those parts of human living which have still not been "swept" into his models? Or do we have to say that the basic aspects of human values have never been and never can be represented by the approaches to systems that the management scientist, the planner, and the behavioral scientist adopt?

The richest thing I think we can gain from the discussions of anti-planning is the understanding of the really basic conflict. In the process of preparing proposals, conducting research, and writing out recommendations, the management scientist and the planner are apt to become convinced that their approach to systems is the correct approach. They are supported in this idea if the manager or the politician goes along with them. They are further supported if the recommendation is implemented, and they see their freeways, medical systems, and educational systems in actual existence. What they forget, of course, is that unseen, unheard part of humanity which has never entered into the domain of their vision or thinking. What they fail to see in their detailed analysis of cost benefits is that the system they have created may be largely irrelevant or perhaps even partially destructive for the person who finds his life in the religious, or in the search for the self, or in the completely nonintellectual.

There is something fiercely ugly at times about proposal writing, testing, and implementation of the programs of large-scale systems. It's much as though the whole life of the system had been depleted in an attempt to put it into a rational mode.

In any event, no discussion of the systems approach should omit the viewpoint of the anti-planners, who regard the planners as intruders and snoopers—snoopers, because, using the methods of behavioral scientists, they come into a person's life to steal information from him; intruders, because they believe it perfectly appropriate to interrupt or change the pattern of one's living without even so much as a "please." Furthermore, if there is a bit of conscience among the planners, they feel perfectly satisfied if the persons interviewed or planned for say that they are willing to have the interview or the plan made. To the anti-planner, this willingness on the part of the person in no way excuses the unforgivable behavior of many a planner and behavioral scientist.

Thus, anti-planning must essentially be regarded as a fundamental part of the systems approach. No approach to systems can stand by itself. Its only method of standing is to face its most severe opposition.

14. CONCLUSION

It is impossible to write a chapter which concludes a debate of the kind that has been carried on in this book. There can be no conclusion. The best that can be done in a final chapter is to say something about this raconteur, this third party, who sat aside and felt perfectly free to discuss various approaches to the system as though he himself were above suspicion. Who is this author, after all? Is he really the scientist or planner? Or the anti-planner? Or what? Which side is he on? Where does he stand? Is he free from suspicion and criticism simply because he looks at all of these activities from many different viewpoints? Is his the supersystem approach?

When I began to write this book on a request from the publisher, I thought of it more or less as a popular text on the systems approach in which I would discuss many of the scientist's techniques and methods. But as I started to write in earnest, I began to see how difficult it was simply to describe to the reader how the management scientist behaves and persuade him that this behavior has some real benefit. In a way the very writing of the book forced me into the debate. The only tolerable way to write a book of this kind was to inject the criticism into the very context in which a technique was being discussed. Indeed, if I were to think of one theme that has been in the back of my mind as I wrote these chapters, it is the theme of deception. You see, the management scientist at the outset felt that the efficiency expert was deceived. The efficiency expert, he said, believes that, when he sees idleness and slack in the system, he is looking at a reality. From the management scientist's point of view, he is looking at an illusion. He is tricked by his perceptions. But then the management scientist, when he becomes very serious about his own models, in which "all" of the objectives are represented and a "proper" compromise is created, also is deceived. In the straight-faced seriousness of his approach, he forgets many things: basic human values and his own inability really to understand all aspects of the system, and especially its politics.

I came to this notion of deception in a brief experience with extrasensory perception. I was amazed to see how many psychologists had taken extrasensory perception so seriously. Extrasensory perception is a way of looking at the world, a world view in which some human beings have an ability of receiving messages from the future or from distant places or whatever, without the use of the ordinary sensory apparatus. Those who believe in extrasensory perception are deadly serious about it. They keep their face straight because one crack of a smile would indicate a weakness on their part and lay them open to the even more severe attacks from their enemies. But what was also interesting was the deadly seriousness of their opponents. Extrasensory perception apparently was no joke. If it succeeded, it would destroy the basis of psychology, and perhaps even of physical science. Hence, although it could not be taken seriously, it could not be taken humorously either. In both cases, deception, it seemed to me, was rampant. Perhaps the believer in extrasensory perception, in his severe insistence on the existence of mysterious contacts with reality, is deceived; but so also is the serious-minded scientist who is completely convinced that there is only one way to look at reality, namely through the recognized sensory channels that have been laid out in the foundations of psychology.

Carrying over this experience of extrasensory perception to the systems approach, I arrive at the conclusion that however a systems problem is solved—by planner, scientist, politician, anti-planner, or whomever—the solution is wrong, even dangerously wrong. There is bound to be deception in any approach to the system.

And yet when one looks at the solution and sees its wrongness, one is also deceived, because, in searching for the wrongness, one misses the progressive aspect of the solution. We have to say that the advocate of the solution both deceives and perceives. We have to say that the solution is ridiculous and serious. We have to maintain the contradiction or else we allow ourselves to be overwhelmed by the consistent.

And so in the end I come to science, which has been the main topic of conversation in the entire book. At no one point did I stop to define for the reader what science means, although once or twice I characterized it in terms of observation and reasoning. I think it's deceptive to look at science solely as a series of activities carried on by people who call themselves scientists. Science itself is a system subject to considerable change, as we have seen in the last few centuries. It's very deceptive to believe that science has arrived at a plateau where its own change is minimized. Instead, the science of our society has to be looked at as a system itself subject to change. If the anti-planner really believes that he has arrived at a truth about himself or about the way in which God governs the systems of the world, then the anti-planner as a scientist may be a deceived scientist, just as he believes that the management scientist is deceived.

The ultimate meaning of the systems approach, therefore, lies in the creation of a theory of deception and in a fuller understanding of the ways in which the human being can be deceived about his world and in an interaction between these different viewpoints.

As I have been writing this book, the United States has been waging a serious war in Southeast Asia. Is there a systems approach to warfare? Not in the minds of most observers and participants. The hawks want to win, to "stop aggression," to "support our national policy." They can't look at the situation in any way except their own way. They regard peace demonstrations as "support of the enemy." The doves say the war is ridiculous, that we must "pull out"; they can't look at the situation in any way but their own way.

In the beginning I listed some things the world could very well afford to do: feed and clothe its poor, for example. But each person looks at this problem in such a one-sided way that the systems approach is lost.

Hence, I, too, am biased and deceived. It's naïve to think that one can really open up for full discussion the various approaches to systems. People are not apt to wish to explore problems in depth with their antagonists. Above all, they are not apt to take on the burden of really believing that their antagonist may be right. That's simply not in the nature of the human being.

Well, then, what is the systems approach? On the one hand, we must recognize it to be the most critical problem we face today, the understanding of the systems in which we live. On the other hand, however, we must admit that the problem—the appropriate approach to systems—is not solved, but this is a very mild way of putting the matter. This is not an unsolved problem in the sense in which certain famous mathematical problems are unsolved. It's not as though we can expect that next year or a decade from now someone will find the correct systems approach and all deception will disappear. This, in my opinion, is not in the nature of systems. What is in the nature of systems is a continuing perception and deception, a continuing re-viewing of the world, of the whole system, and of its components. The essence of the systems approach, therefore, is confusion as well as enlightenment. The two are inseparable aspects of human living.

Finally, then, here are some principles of a deception-perception approach to systems:

1. *The systems approach begins when first you see the world through the eyes of another.*

Another way to say the same thing is that the systems approach begins with philosophy, because philosophy is the opportunity to see the world through the eyes a Plato, a Leibniz, or a Kant. The reading of philosophy is not an abstract study; the serious student takes on the burden of becoming convinced that each important philosophical position is right, absolutely right. He relives the intellectual vitality of the past. He feels to the utmost that the real world is the modeled world; that the real world is the experienced world; that the real world is dialectical; and so on. He does all this without losing his own individuality.

2. *The systems approach goes on to discovering that every world view is terribly restricted.*

That is, every "world view" looks only at a component of some other system. For those who think in the large, the "world" is forever expanding; for those who think in the small, the inner world is forever contracting.

3. *There are no experts in the systems approach.*

As I was writing this final chapter, I turned on the TV one Sunday for relaxation. There were a Catholic priest and an Episcopalian minister discussing the "new morality." The priest was saying that many people believe now in "situational ethics," doing what seems right to you at the moment. The minister replied that he knew of no reputable theologian who took such an extreme view. The priest looked startled; *he'd* thought that the "new" morality referred to the younger generation and their older admirers in the public, and not to the opinions among the experts. He was right, of course. The real expert is still Everyman, stupid, humorous, serious, and comprehensive all at the same time. The public always knows more than any of the "experts," be they economists, behavioral scientists, or whoever; the problem of the systems approach is to learn what "everybody" knows.

And finally, my bias:

4. *The systems approach is not a bad idea.*

Supplementary Sections

SUPPLEMENT I

SOME EXERCISES IN SYSTEMS THINKING

1. Describe the following systems, if you are familiar with them, by identifying the components, resources, etc.

> urban rapid-transit system
> grammar-school system
> laundry system
> grocery-store system
> your own household

Now develop a five-year plan for the system you have described, following the steps given in Chapter X. Where necessary, indicate the lack of information and evaluate the benefit of obtaining additional information.

2. Imagine that, because of some quirk in man's inventive capabilities, no faster or more convenient mode of transportation

had occurred up to now than the horse and carriage, though all other technologies were up to date, including communication. Now a bright young engineer discovers the combustion engine. Granted an excellent ability to forecast consequences of the type we so well know today (accident, speed, smog, convenience), what is a good approach to the "new" transportation system (including making the engine illegal)? How does this exercise help in understanding today's transportation problems, if at all?

3. Given that the computers of 20 years from now will be small, cheap per unit of computation, and extremely fast, what are the implications for (*a*) library systems; (*b*) newspapers; (*c*) advertising?

4. Contrast the systems approach and the "privacy approach." Do you see how a systems approach to the planning of large cities runs counter to the human need for privacy? "Privacy" in this question means dignity and respect, independent of income and status. Complete privacy also means that the individual is not investigated and information about him is not recorded, *even though he himself may not object.*

5. Pick out your favorite war and ask yourself which side really had the correct systems approach, or, if that question annoys you, whether in war it is ever possible for any "side" to have the systems approach. If you feel annoyed, does the feeling convert you into a pacifist? (What is the meaning of "peace" in the systems approach)?

6. Nowadays we hear a lot about "underdeveloped" countries. What would the measure of performance of a nation have to be for the United States to be the most developed country in the world today? The least developed country? Is such a use of "measure of performance" appropriate (i.e., in which one asks what the real world would have to be like for an existing system to have the highest—or lowest—score)?

7. A new automobile costs 3,000 dollars. It costs 500 dollars to run it the first year, in addition to 200 dollars in interest charges on the loan. Each subsequent year it costs 200 dollars more to run the car (including the cost of its "appearance"),

while the interest charges go down 50 dollars a year. The salvage value is 2,000 dollars at the end of the first year, and 400 dollars less each subsequent year (1,600 dollars second year, 1,200 dollars third year, etc.) When should you buy a new automobile, assuming that your credit is always good? Does solving this exercise help you to decide when *you* should buy another automobile?

8. In what sense do the following slogans succeed or fail to capture the systems approach: (1) "Support your local police"; (2) "Stem the tide of Communism"; (3) "Workers of the world, unite!" (4) "Don't be a litterbug."

9. Which political ideology, if any, has the most thoroughgoing systems approach: right wing, middle of the road, liberal, left wing, other?

10. A young man is faced with the choice of taking one of two job offers. He decides to approach the choice systematically. First, he writes down his objectives, in the order of their importance to him, as follows: salary; promotion opportunity; pleasant climate; closeness of friends (these aren't his only objectives in life, but the rest were equally well served by both jobs, so he leaves them out). He finds that job *A* serves the highest-ranking objective (salary) best, but is worse than job *B* with respect to the other objectives. Hence, he sees he has to go further and quantify his preferences. He assigns 1 to "salary" and a number less than 1 that expresses his relative preference for the others. The result is:

salary	1.0
promotion	0.8
climate	0.2
friends	0.1

Now he realizes something is wrong, because he does not really prefer the higher salary of job *A* to the combined promotion opportunity and climate of job *B*, and hence the numbers don't add correctly (.6 + .2 is less than 1). So he revises his estimates to make the results consistent:

salary	1.0
promotion	0.8
climate	0.3
friends	0.2

Now he is satisfied with the way the numbers add up. Next he realizes there is a probability associated with each choice; for example, it is possible that job *A* would come through with a promotion sooner than he expected. So he considers each choice against each objective and assigns subjective probabilities to them. The results are as follows:

	Probability		*Probability*
Job *A* and salary	1.0	Job *B* and salary	0
Job A and promotion	0.2	Job *B* and promotion	0.7
Job *A* and climate	0	Job *B* and climate	1.0
Job *A* and friends	0.2	Job *B* and friends	1.0

He then scores each job by multiplying the probability by the final preference numbers thus:

$$\text{Job } A = (1.0)(1.0)+(0.2)(0.8)+(0)(0.3)+(0.2)(0.2) = 1.20$$
$$\text{Job } B = (0)(1.0)+(0.7)(0.8)+(1.0)(0.3)+(1.0)(0.2) = 1.06$$

Consequently, he decides on job *A*. Criticize this technique (would you use it, why it might be deceptive, etc.?)

SUPPLEMENT II

SUGGESTED READINGS

To read more about the systems approach, one might begin by reading some philosophy, to see how up to date the past really is. Plato's *Republic* is a famous systems-science book, wherein the author attempts to design the political structure of a state and thence makes some inferences about justice, a very important topic in systems science. The *Republic* is relatively easy reading and has the advantage of being thoroughly irritating at times, which every good book should be. Aristotle's *Ethics* is a bit harder and perhaps is not really a "systems" book at all, but his "mean between two extremes" is an early version of the management scientist's more precise compromise between conflicting objectives; e.g., the efficiency expert goes to one extreme, the affluent spender to the other, and the "optimum" is the "mean between the two extremes," a weighted mean to be sure.

The "pre-Socratics" are even fresher than Plato and Aristotle, and are mainly interested in the "whole system" without the bias that the accumulations of science and history have produced. Nietzsche's *Philosophy in the Tragic Age of the Greeks* gives some of the flavor.

Needless to say, St. Thomas Aquinas' *Summa theologica* is a monumental work on the systems approach, but for the most part the medieval ideas about whole systems were taken up by later writers in the modern period.

In modern times there is Thomas Hobbes' *Leviathan,* an account of the origin and structure of human societies, and René Descartes' *Discourse on Method,* which gives one of the clearest arguments for the need for a system guarantor. Leibniz's *Monadology* is harder reading, but a very astute account of what the "whole system" may be like. Spinoza's whole system is described in his *Ethics,* for advanced readers; his *Treatise on the Improvement of the Understanding,* however, is one of the best "how to think" books ever written. Jeremy Bentham's *Introduction to Principles of Morals and Legislation* is one of the earliest and probably the best-considered programs for what I have called the economist's or planner's approach to social systems. If difficult reading does not trouble you, then also look at Immanuel Kant's *Critique of Practical Reason,* an entirely different approach from Bentham's. For Kant, it is the "moral law within" that ought to determine good and bad conduct; only "in the limit" do Bentham's and Kant's systems approaches become one and the same.

The nineteenth century produced many writers on the nature of whole systems: Hegel, Marx, Schopenhauer, Nietzsche, Spengler, Spencer, to mention a few. The century seems to have been fascinated by the idea of destiny, whether it took the form of Absolute Mind, Dictatorship of the Proletariat, Nihilism, Superman, End of Civilization, or Biological Evolution. The idea of destiny has much to do with the systems approach, because it helps us to conceptualize the elusive problem of future generations. In this century, however, we have become rather indifferent to its significance, because we are far more interested in the feasible, i.e., what we can see, touch, and directly change *now*. Only in very recent years have social thinkers turned to predicting the future of human society. See, for example, B. de Jouvenal's *Art of Conjecture* (N.Y.: Basic Books, 1966), Olaf Helmer's (and others') *Social Technology* (N.Y.: Basic Books, 1966), and Herman Kahn's and Anthony J. Wiener's *The Year 2000* (N.Y.: Macmillan, 1967).

There are many contemporary books and articles on the systems approach and closely related topics. Those I have selected provide a wide variety of viewpoints and, in most instances, cite other material that will broaden the reader's perspective.

For an introduction to management science and operations research, there are *A Manager's Guide to Operations Research*, by R. L. Ackoff and P. Rivett (N.Y.: John Wiley and Sons, 1963); and *Executive Decisions and Operations Research*, by D. W. Miller and M. K. Starr (Englewood Cliffs, N.J.: Prentice-Hall, 1960). *Introduction to Operations Research*, by C. W. Churchman and others (N.Y.: John Wiley and Sons, 1957), is a text, but Part I describes a methodology of the systems approach. *Executive Readings in Management Science*, edited by M. K. Starr (N.Y.: Macmillan), is a collection of nontechnical readings on the significance of management science's approach. A more recent book, *Decision and Control*, by Stafford Beer (N.Y.: John Wiley and Sons, 1967), combines insight with practical experience and emphasizes the use of cybernetics in the systems approach.

Resource Allocation (Vol. III of *Surveys of Economic Theory*, edited by E. A. G. Robinson [N.Y.: St. Martin's Press]), contains a survey article on cost-benefit analysis by A. R. Prest and R. Turvey, as well as articles dealing with the economic approach to systems. The article by J. R. Hicks on "Linear Theory" provides an elaboration of the linear-programming ideas discussed in the text. *Socials Indicators*, edited by R. A. Bauer (Cambridge, Mass.: Massachusetts Institute of Technology Press), discusses the possibility of extending the logic of economic analysis to include "social" benefits.

There are various behavioral-science approaches to systems. Herbert Simon discusses the idea of "satisficing" compared to pure optimizing in *The New Science of Management Decision* (N.Y.: Harper and Row, 1960); his idea is that practicing managers do not seek beyond a "level of aspiration." F. E. Emery and E. L. Trist describe "Socio-Technical Systems" in *Management Sciences,* edited by C. W. Churchman and M. Verhulst (N.Y.: Pergamon Press, 1960); their interest is in the involvement of people in technical change. Discussions of other relationships between operations research and social science appear in *Operations Research and the Social Sciences,* edited by J. R. Lawrence (London: Tavistock Publications, 1964). There is a whole series of publications by behavioral scientists in what is sometimes called the "change" literature. *The Planning of Change: Readings in the Applied Behavioral Sciences,* edited by Warren Bennis and others (N.Y.: Holt, Rinehart and Winston, 1961), is a collection of readings in this area. A very fascinating development in this connection is the "invention" of T-groups, a method of releasing antagonisms and fears, thereby (it is hoped) creating a deeper mutual understanding and trust. D. Braybrooke and C. E. Lindblom discuss the idea of "incrementalism," the carefully chosen distance that the next change should make, in *A Strategy of Decision. Policy Evaluation as a Social Process* (N.Y.: Free Press of Glencoe, 1963). "The Researcher and the Manager," by A. Schainblatt and C. W. Churchman (in *Management Science,* 1965), classifies various ideas about how recommendations for system change should be implemented; this paper was followed by a whole issue in the same volume of *Management Science* to which various authors contributed their ideas. Thomas Cowan discusses the relationship of the systems approach to law in "Decision Theory in Law, Science and Technology" (*Rutgers Law Review,* Vol. 17, 1963, p. 499; and *Science,* June 7, 1963).

There are a number of books that discuss the pros and cons

of technological change. *Technology and Change*, by D. A. Schon (N.Y.: Delacorte Press, 1967), is insightful and provocative. *The Technological Society*, by Jacques Ellul (N.Y.: Alfred Knopf, 1964), gives a very pessimistic account of how "technique" is dominating our lives. *Understanding Media*, by Marshall McLuhan (N.Y.: McGraw-Hill, 1964), is a more spirited and popular account of the human being in a technological environment.

An elaboration of the theme of the last chapter is contained in my *Challenge to Reason* (N.Y.: McGraw-Hill, 1968).

Some of the material discussed in this book can be read in greater depth in the following articles and books:

1. The study of the loading and unloading system of a port is reported in *San Francisco Port Study*, Vol. II, Part V, National Academy of Sciences–National Research Council.

2. The information system for the State of California, as well as three other system studies—in crime, sanitation, and transportation—are described in Dr. Ida Hoos's paper, "A Critique in the Application of Systems Analysis to Social Problems," Internal Working Paper #61, April, 1967, Social Sciences Project, Space Sciences Laboratory, University of California, Berkeley. Dr. Hoos's work describes some of the political reaction to these studies.

3. The alcoholism mission is described by A. Schainblatt in "Planning, Programming and Budgeting Applied to the Problem of Alcoholism," 66 TMP-111, December 1966, TEMPO, General Electric Company, Santa Barbara, Calif.

4. Conflict of values in a "game" setting ("prisoner's dilemma") is described in A. Rapoport's *Fights, Games and Debates* (Ann Arbor, Mich.: University of Michigan Press, 1960).